Science Communication Skills for Journalists

Science Communication Skills for Journalists

A Resource Book for Universities in Africa

Written and edited by Dr Charles Wendo

With contributions from Dr Abraham Kiprop Mulwo, Dr Aisha Sembatya Nakiwala, Dr Darius Mukiza, Dr Samuel George Okech and Dr William Tayeebwa

CABI is a trading name of CAB International

CABI
Nosworthy Way
Wallingford
Oxfordshire OX10 8DE
UK

Tel: +44 (0)1491 832111
E-mail: info@cabi.org
Website: www.cabi.org

CABI
WeWork
One Lincoln St
24th Floor
Boston, MA 02111
USA

Tel: +1 (617)682-9015
E-mail: cabi-nao@cabi.org

A catalogue record for this book is available from the British Library, London, UK.

References to Internet websites (URLs) were accurate at the time of writing.

ISBN-13: 9781800621428 (paperback)
9781789249668 (OA ePDF)
9781789249675 (OA ePub)

DOI: 10.1079/9781789249675.0000

Commissioning Editor: David Hemming
Editorial Assistant: Lauren Davies
Production Editor: James Bishop

Typeset by SPi, Pondicherry, India
Printed and bound in the USA by Integrated Books Integrated, Dulles, Virginia

Contents

Contributors

Mukiza, Darius
School of Journalism and Mass Communication, University of Dar es Salaam
Address: University of Dar es Salaam, PO Box 4067, Dar es Salaam, Tanzania
Email: rugalama@gmail.com
Mobile: +255 755 277711

Mulwo, Abraham Kiprop
School of Information Sciences, Moi University
Address: Moi University, Main Campus, PO Box 3900, Eldoret – 30100, Kenya
Tel: +254 713 201327

Nakiwala, Aisha Sembatya
Department of Journalism and Communication, Makerere University
Address: Makerere University, PO Box 7062, Kampala, Uganda
Email: nakiwala@yahoo.com
+256 772 516078

Okech, Samuel George
Department of Veterinary Pharmacy, Clinical and Comparative Medicine, College of Veterinary Medicine, Animal Resources and Bio-security, Makerere University
Address: Makerere University, PO Box 7062, Kampala, Uganda
Email: blessedsgo@gmail.com
+256 772 605586

Wendo, Charles
CABI
Address: CABI Uganda office, NARO Secretariat, Plot 3, Lugard Avenue, Entebbe PO Box 295, Entebbe, Uganda
charles.wendo@scidev.net
Tel. +256 772 421485

Tayeebwa, William
Department of Journalism and Communication, Makerere University
Address: Makerere University, PO Box 7062, Kampala, Uganda
Email: wtayebwa@gmail.com
Tel +256 776 482 892

Foreword

Empowering learners to practise science journalism
Professor Suleiman Bala Mohammed, Vice-Chancellor, Nasarawa State University, Nigeria

In 2018, when Nasarawa State University Keffi (NSUK) was approached by the Science for Development Network (SciDev.Net) to be part of a novel programme called Script that aimed to institutionalize science communication in the university, we did not need to think twice about getting on board. Effective communication is a challenge in many spheres of society in Africa, due in part to the lack of communications infrastructure and the fragmented linguistic landscape. Throw science into the communication mix and the challenge of delivering information to the audiences that need it – and, crucially, in a form that they can understand – is even greater. Little science news is reported in mainstream media and what there is can be lacking in accuracy and scientific rigour. Very few African universities offer courses or programmes in science communication, and African perspectives are largely missing in textbooks on science communication.

At NSUK the Script programme provided us with the opportunity and impetus to meet an already identified need for expert training in science communication for both scientists and journalists. For NSUK this is entirely in keeping with our stated mission of providing "Knowledge for Development", as a way to build strength and depth in science reporting so that scientific knowledge that has been generated can be placed in the hands of those that need it, for the benefit of all society. Since the programme began in 2018, NSUK has embraced the concept and delivery of Script (through, for example, the creation of science communication clubs and awards) and is now at the forefront of driving academic science communication training in Nigeria, acting as an advocate within the academic community and sharing our knowledge, skills and experience with several other universities in the country.

This book is therefore timely. Its backbone is tried and tested Script materials that have already been used to successfully train thousands of students in science communication. Given that the Script training has, at the time of writing, already led to 46 graduates securing science journalism internship opportunities in media institutions in Nigeria, I am confident that student users of this great resource will be empowered to practise science journalism. The book's approach to learning and practising science communication makes it both pedagogically sound and interesting to read and understand, with a mix of theory and practice that ensures that learners understand the principles which underpin science communication prior to applying them. The discussion questions and suggested answers make it useful for both teachers and learners.

The book recognizes that for effective science communication to take place, journalists need scientists and scientists need journalists. It values interdisciplinarity: journalism students are encouraged to think like scientists while, conversely, science students or scientists are asked to think like journalists. This unique approach encourages cross-disciplinary interactions, bridging the gap between scientists and journalists, and building trust among these professions.

I highly recommend this book to any African university delivering a science journalism and communication course, and to any student who wants to better understand the need for and practice of communicating science.

Foreword

The significance of effectively communicating science
Professor Barnabas Nawangwe, Vice-Chancellor, Makerere University

Universities and research institutions worldwide continue to undertake profound research in various scientific disciplines. But often these institutions fail to prioritize the dissemination of the knowledge this research yields to non-scientific audiences, such as members of the public and policymakers. As a result, they lose out on opportunities for their research to influence policy and shape the national conversation; and, increasingly, the potential to access future funding opportunities, as donors are placing increasing importance on research communication. Effective dissemination of science is essential to ensure that research findings inform policies to help address societal challenges.

It is in this spirit, and in acknowledgement of this challenge, that Makerere University partnered with the Science for Development Network (SciDev.Net) to embed science journalism and communication in the university's curricula through the Script programme. This collaboration has mainstreamed science journalism and communication in our university's undergraduate and graduate curricula, encouraged interactions between students in journalism and communication and their counterparts in the natural, biological and physical sciences, and linked faculty and students to industry professionals in science communication. The collaboration led to the creation and subsequent approval of three new stand-alone courses with a focus on science journalism and communication – Science Communication; Health and Environment Communication; and Information, Communication and Knowledge Management – as part of the Master's Degree in Strategic Communication programme.

One of the challenges that this work presented was the need to revamp the communication and journalism curricula in order for them to incorporate, and be aligned with, best practices in science journalism. SciDev.Net, which has experience in practical science journalism, provided the needed resources and expertise to do this. What a good thing it is to see that part of this expertise can now be found in this book, *Science Communication Skills for Journalists: A Resource Book for Universities in Africa*.

This book will undoubtedly help African universities that want to start new science journalism and communication courses or programmes, since they now have a book authored by experts in the subject matter who have experience on the continent. The book also closes the gap between academic institutions and industry, thus ensuring that users will find it easier to work with media institutions that highlight science journalism and communication. Indeed, the book challenges media institutions to place science in African newsrooms, and offers reasons for doing so. The book clearly demonstrates SciDev.Net's expertise in acting as a bridge between key stakeholders such as scientists and policymakers to bring about development progress. I strongly recommend this great resource to African institutions that value both science and communication.

Preface

A new dawn for science journalism in Africa

Dr Charles Wendo, Training Coordinator, SciDev.Net, CABI

I came into a science journalism career by chance three months after I graduated with a bachelor's degree in Veterinary Medicine. At that time, Lake Victoria, Africa's largest freshwater body, was choking with water hyacinth. The dreaded water weed formed expansive green mats all over the lake, covering 80% of the shoreline and hampering fisheries, water transport and beachgoing. It had become a significant problem for several countries.

In this context, during my final year at university I planned to conduct a research study to determine whether water hyacinth might have a positive use – as livestock feed. I read similar research studies carried out in Asia and South America. However, it turned out that the study was too expensive for me to conduct at the time, so I had to abandon it. I put away the literature I had gathered about water hyacinth and instead carried out a more affordable research study to fulfil my degree award requirements.

Three months after the graduation ceremony it occurred to me that the public and policy-makers needed to know what I knew about water hyacinth, so I dug up the literature I had amassed and wrote a feature article for *New Vision*, Uganda's leading newspaper. In the report I told the paper's readers that water hyacinth might be able to be used to feed livestock. I quoted research studies from other parts of the world and I argued that similar studies needed to be carried out in Uganda.

The editors at the newspaper liked the article because it was topical, it referred to credible scientific evidence and it was written in simple English that most readers would understand. The editors encouraged me to continue writing popular science-based articles. After writing part-time for two years, I took a full-time job at the newspaper. I thus learnt my journalism skills through on-the-job training. Later, I went back to university to obtain a postgraduate diploma in Mass Communication and then a master's degree in Journalism and Communication.

That was then. Today, you don't have to go through a long and tortuous journey to become a science journalist in Africa. There are two main ways to become a science journalist nowadays: either be a scientist who learns journalism or be a journalist who learns science. Regarding the latter route, journalism students in several African universities now have the opportunity to learn the skills they need to report science stories. Between 2019 and 2021, at least four African universities introduced science journalism and communication into their undergraduate and postgraduate curricula with the support of SciDev.Net's science journalism and communication capacity building programme, Script. Several other universities have introduced science communication into their curricula without the support of the Script programme.

Despite the progress signalled by the availability of university courses on science journalism, both lecturers and students complain that there is a lack of resource books for science journalism training in Africa. The Script programme aims to fill this gap with the present book, *Science Communication Skills for Journalists: A Resource Book for Universities in Africa*. Learning institutions can use the book as reference material for science journalism training. The book aims to be helpful to a wide variety of readers: university students, lecturers, practising journalists and media trainers.

The book is divided into two parts. Part I is made up of six chapters which present academic discussions on the practice of science journalism. Part II is made up of ten chapters which focus on the skills needed for science journalism, with hands-on advice, examples and learning activities.

The context for the publication of this book and other activities of the Script programme is the continued need to increase the quantity and quality of science stories reported in the African media. Scientific research provides the knowledge that policymakers and the public use to make informed decisions. Policymakers and the public rely on the media as their primary source of information on scientific developments that are relevant to them. In this regard, we need the media to report science continually. However, to date, Africa has faced a shortage of journalists with the competence to report on science.

At the same time, it should be stressed that this book comes at a time when the outlook for science journalism in Africa is quite positive. For example, the Global Science Journalism Survey 2021 showed that sub-Saharan Africa had younger and happier science journalists than most other regions of the world. And a UNESCO report, *Media Coverage of Science and Technology in Africa*, has indicated that the relationship between scientists and journalists in Africa is increasingly symbiotic.

Many months of consultation, commissioning, writing, editing and publishing have led to the book you are now reading. We hope it will be a valuable resource, whether you are a student, a lecturer or a journalist, and we hope it will contribute to the growth of science journalism in sub-Saharan Africa.

We wish you a fruitful reading!

Acknowledgements

We are grateful to Robert Bosch Stiftung (RBS) for the financial support towards the production of this book and other activities of the Script programme. Special thanks go to Isabella Kessel at RBS for her constant guidance and support to the Script team. Our gratitude also goes to the contributors: Abraham Kiprop Mulwo, Aisha Sembatya Nakiwala, Darius Mukiza, Samuel Okech and William Tayeebwa. We appreciate the academic staff at Nasarawa State University Keffi in Nigeria, Makerere University in Uganda, Moi University in Kenya and University of Dar es Salaam in Tanzania, who tested out the content of this book in class and provided valuable feedback. A big thank you to all those who participated, in one way or another, in planning the book and casting an eye over its contents, including Paul Rogers, Paul Dawson, Sarah Hebbes, Bookie Ezeomah and Juliet Tumeo. Finally, special thanks to the team at My Blue Pencil for proofreading the manuscript and the CABI publishing team for preparing and producing the book.

Part I

Theory and Experience

This part presents the theoretical concepts underpinning the need for, and practice of, science journalism and communication in Africa. Of the six chapters, five present academic discussions on the practice of science journalism and communication in Africa. One chapter presents a case study detailing the experience of Uganda's Makerere University in introducing science journalism and communication into its undergraduate and postgraduate curricula.

This part of the book sets the foundation for Part II, which focuses on practical skills for science journalism.

What is Scientific Research and How is it Conducted?

Dr Charles Wendo
Training Coordinator, SciDev.Net, CABI

In this chapter
- What is scientific research?
- How scientific research is conducted
- Uses of scientific research
- Branches of science
- Types of scientific research
- What constitutes scientific evidence?
- Linking scientific research and journalism
- Similarities and differences between scientists and journalists

---

What is scientific research?

It is important for anyone intending to report science news to have some basic knowledge about what scientific research is, how it is done and how it fits with the news media. This chapter begins by discussing the definition of scientific research, describing the process of conducting such research, and highlighting the application of scientific research in general. It goes on to discuss the different types of scientific research and how to interpret them. Also included is a discussion about what constitutes scientific evidence, and how to make sense of it. Finally, the chapter introduces the reader to the link between science and the media.

Why is the sky blue? Why are most flowers brightly coloured? Why is it hotter in some months than others? Where does rain come from? Why do rocks have different colours?

Science has answers to these and many other questions about the world around us. But how do scientists know the answers to these questions? They know because the world around us has patterns, which have been studied and documented for many years.

Before something is accepted and documented as scientific knowledge, a systematic process of testing ideas takes place, in order to prove it. This process is called scientific research. Thus, scientific research – also known as the scientific process – is how scientific knowledge is discovered. Thus, carrying out scientific research may also be referred to as "doing science".

Science is not static: scientists never stop carrying out research to discover new knowledge in various specialities to answer various questions, to add to the body of scientific knowledge and to address societal concerns. For instance, in 2019 scientists announced the discovery of an Ebola vaccine and two effective therapeutic treatments for the disease. Subsequently, the World Health Organization approved the vaccine for use in the prevention and control of Ebola virus disease (World Health Organization, 2019). In fact, the amount of scientific research carried out around the world is growing every year: for instance, there was a 23% increase in the number of published scientific research articles between 2008 and 2014 (UNESCO, 2015, p. 37).

How scientific research is conducted

Scientists carry out research using a process known as the scientific method, to discover and document new knowledge. The process begins with a scientist observing the world around them, asking important questions about what they observe and proposing a tentative answer, referred to as a hypothesis. This is followed by a systematic process of collecting and analysing data to find the real answers (Fig. 1.1).

A good example of the scientific process is the discovery of gravity by Sir Isaac Newton in 1666. While sitting under an apple tree, Newton saw an apple fall down (Stukeley and White, 1752, p. 15). He wondered why it fell downwards. Why did it not go sideways or upwards or just remain in the air? He concluded that there must be some force pulling the apple towards the earth. Newton then carried out a series of investigations that led to the discovery of gravity. To this day, Newton's law of universal gravitation is taught in schools. In the same way as Newton's discovery of gravity, most researchers begin by noticing an issue and raising relevant questions about it.

Scientific researchers collect data by making observations, taking measurements, asking human beings to answer certain questions, carrying out experiments, examining existing databases or reviewing documents. They then analyse this data, interpret the results and make conclusions, as well as recommendations.

Example: Researching the cause of cholera

Until the 1850s, medical doctors generally believed cholera was caused by bad air. However, Dr John Snow, a physician from London, thought otherwise (Frerichs, 2009; Lippia and Gotuzzob,

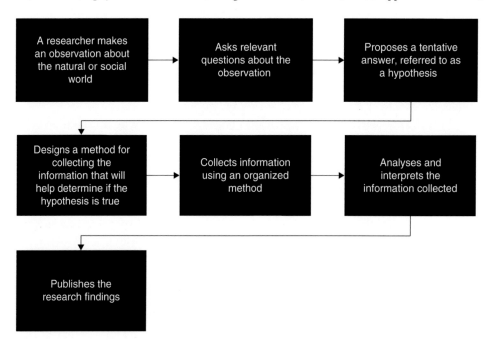

Fig. 1.1. Diagrammatic representation of how scientific research is conducted.

2014). He wondered why the main symptoms – diarrhoea and vomiting – occurred within the digestive system. He suspected the disease might be caused by something that human beings consume in food or water. In an attempt to prove his point, during a cholera outbreak in the early 1850s, Snow interviewed affected families and plotted their homes on a map. This process enabled him to establish that most of the cholera cases were linked to a particular water pump. Additionally, he examined the water from that pump under a microscope and saw some whitish particles that were not present in water from other sources. Armed with this new evidence, Snow convinced the local authorities to shut down the suspicious pump. This ended the cholera epidemic. However, this was not enough evidence to convince the medical community that the cholera epidemic was indeed caused by germs that people had consumed in contaminated water. Years later, other scientists, such as Filippo Pacini and Robert Koch, eventually proved what Snow had suspected – that cholera is caused by certain microscopic organisms. Their discovery eventually led to the germ theory of disease, which states that many diseases are caused by micro-organisms.

Uses of scientific research

When packaging information about scientific research for a wider audience, it is important to have in mind who is likely to use the information, and for what purpose. Thinking about how research findings may be used can help a journalist or communication specialist to pitch the information appropriately.

The knowledge generated through scientific research can be used in different ways.

- Policymakers use knowledge generated through scientific research to make better policy decisions for society. For example, if a study shows that climate change has become a major driver of migration, governments may consider prioritizing climate change action.
- Individuals use knowledge generated through scientific research to make decisions about how to improve their lives, such as exercising daily if research shows that those who exercise more live longer.
- Professionals, such as doctors, engineers and agriculturalists, use knowledge generated through scientific research to improve their work. For example, a study on a faster and cheaper test for a virus can help doctors manage a disease better.
- Researchers use knowledge generated through scientific research as a starting point for additional research: for example, the identification of a new potential biological weapon against the fall armyworm can be used by another scientist to develop a tool to control the pest.
- Industries use knowledge generated through scientific research to manufacture a new line of products. For example, if a study shows that a new method of testing malaria without drawing blood is effective, this information can be used to produce and commercialize a blood-less malaria testing kit.
- Teachers use knowledge generated through new scientific research to update their curricula and replace outdated scientific theories with current thinking.

Branches of science

The natural sciences involve the study of the physical world. The key word is nature. This includes basic sciences like biology, chemistry, physics and geology.

The applied sciences involve putting scientific knowledge to practical use to solve human problems. The applied sciences include medicine, engineering and information technology.

The social sciences involve the study of human to human relationships as well as the interactions between people and the natural world. The key word is society. The social sciences include various branches of knowledge, such as economics, sociology, psychology, political science, law, history, linguistics and anthropology.

To give an example: a study of changes in the quality of water in Lake Victoria over time, and how it is affected by settlement patterns around the lake, is essentially a natural sciences study; on the other hand, research on how and why people settle around the lake is a social science study. However, the gap between the natural and social sciences is narrowing. Increasingly, scientists are carrying out interdisciplinary or transdisciplinary research that involves both the natural and social sciences: for example, research into factors that influence farmers' adoption of new crop varieties.

Types of scientific research

Scientific papers and presentations usually have a "methods" section that explains how the study was carried out. There are different ways to categorize research, based on the method used. These include:

- basic vs applied research
- experimental vs non-experimental research
- descriptive vs correlational vs explanatory
- qualitative vs quantitative
- mixed methods research.

It is important for a journalist to take note of the type of research they are reporting on because it determines how the journalist should interpret and report the findings correctly. Science stories can be useful to society only if the science is interpreted and reported correctly.

Basic vs applied research

Basic research is carried out to generate more knowledge about a phenomenon, but may not have immediate application within society. For example, a researcher may discover a particular substance that occurs in the breath of a malaria patient; this knowledge may not be immediately applicable to society but other researchers can then use it to develop a bloodless test for malaria that is conducted by analysing the patient's breath. When reporting a story about the findings of basic research, these findings should be put in perspective, to make clear to people that the findings do not offer an immediate solution to a problem.

Applied research, on the other hand, addresses a practical problem, such as Dr John Snow's cholera study referred to earlier. The results may be used immediately to inform professional, policy or personal decisions. For example, a study in Morocco has established that a simple filter can make domestic waste water clean enough for use in irrigation (Vesper, 2017). Policymakers, farmers and engineers can utilize this information immediately. A journalist reporting on this study can make the story more meaningful by providing their audience with information on the practical application of the study's findings.

Experimental vs non-experimental research

Experimental research

In a typical experiment, groups of people, animals, plants or objects are given different treatments, or interventions, and the results are compared. For example, a crop scientist may want to know whether changing to a new type of fertilizer leads to higher yields. An experiment enables the researcher to prove that one thing causes or leads to something else. Or it could show that one thing is better than another. To prove that changing to a new fertilizer leads to higher crop yields, a scientist has to grow two or more sets of plants under similar conditions, with the only difference being the fertilizer used. Because experimental research can prove causation, journalists can report confidently that one thing causes another.

To give an example: in July 2019, a team of researchers at the University of Pennsylvania in the USA reported the results of an experiment that found that a shorter course of radiotherapy treatment for prostate cancer patients was as effective as the standard treatment (Kishan *et al.*, 2019). Some patients were given the standard 44-dose treatment while others received a shortened 28-dose treatment. By comparing the two categories of patients and their rate of improvement, the scientists established that the shortened treatment was as effective as the standard. This is a typical example of experimental research.

Non-experimental research

It is not always possible to carry out an experiment, especially where the study involves human beings. For instance, in 2018 the BBC reported a story titled "Men with low sperm counts at increased risk of illness" (Therrien, 2018). It would be unthinkable to experimentally lower or increase men's sperm counts to see if it increased or reduced the risk of other diseases. A more realistic approach is to find men who already have a low sperm count and test them for other diseases, which is what the scientists did in this case. Such men were found to have more body fat, higher blood pressure and more "bad" cholesterol. This was a non-experimental study because it did not involve an experiment. When reporting on non-experimental studies, journalists should exercise caution: they need to bear in mind that just because two things happen together, this does not mean that one causes the other. The fact that men with more body fat have a lower sperm count does not mean body fat reduces sperm count, or vice versa – there could be another factor that increases body fat, while reducing the sperm count concurrently. That is why in this case the researchers concluded that men with a low sperm count should also be tested for other possible health problems.

Descriptive vs correlational vs explanatory research

Descriptive studies

Descriptive studies use numbers to describe patterns. If, for instance, a new crop disease is identified, research can be conducted to establish how prevalent the disease is, where it is most concentrated, which varieties of crop it affects most and when it occurs most. For example, the story "Africa and Asia lead in proportion of blind adults", published by SciDev.Net (Achieng, 2017), is about a study that sought to establish patterns of blindness in adults. The researchers established that the prevalence of adult blindness was higher among women and people with low- and middle-income status. When reporting on descriptive studies, journalists need to be aware that such studies only describe the situation as it is: they do not explain why. If a study only shows that trees of a certain species grow better in one county than another, a journalist should not speculate on why that county is better for that tree.

Correlational studies

Correlational studies look at the relationship between variables, such as between the level of parents' education and child mortality. You need to be cautious in reporting the results of correlational studies. In research, it is often said that "correlation is not causation": if the children of more educated parents live longer, it does not automatically mean parents' education makes children live longer. There could be another factor leading to better education and health concurrently, such as the family's economic status. A correlational study may not necessarily explain why the children of educated parents live longer.

Explanatory studies

These studies explain how two variables are related. For example, the BBC (2018) online story "More evidence essential oils make male breasts develop", published in 2018, refers to a study that attempted to explain a suspected link between essential oils and abnormal breast development among boys. The study found that lavender and tea tree essential oils have eight chemicals that could potentially interfere with male hormones.

Qualitative vs quantitative research

Quantitative research

Quantitative research involves measurements, numbers and comparisons: for example, the percentage of nurses who show signs of stress, and how that compares with people in other professions. To give another example, the article "Child malaria deaths 'slashed by rainy season regimen'", published by SciDev.Net in November 2020 (Nyamol, 2020), is about a quantitative study in which scientists compared the number of malaria cases among children who received periodic antimalarial treatment and those that did not. They found that malaria cases were much fewer in number in communities where children received the treatment, referred to as seasonal malaria chemoprevention (SMC). Quantitative research gives journalists numbers that they can report on, and they should take care to report the numbers accurately. However, numbers alone do not make a good story: a journalist reporting on quantitative research needs to explain the implications of those numbers for society. Additionally, they can make the story more relatable by adding faces to the numbers: for example, by including the experiences of affected people.

Qualitative research

Qualitative research, on the other hand, answers research questions using non-numeric data – frequently text, but also observations, videos and other media. It often focuses on a deeper understanding and description of the issues – for instance, how nurses experience stress – and less on the ability to replicate the results of the study. Where quantitative research addresses the number of people who have access to a safe water source, a qualitative study may focus on the details of how access to safe water affects people's quality of life. Therefore, the methods employed in qualitative research often require detailed engagement with the respondents to collect their perceptions and experiences: for example, in-depth interviews and focus group discussions. It follows that in reporting the findings of qualitative studies, journalists should pay more attention to the description of experiences and observations than to numbers.

Mixed methods research

Scientific researchers often use a combination of both quantitative and qualitative methods. This is referred to as mixed methods, or multiple methods, research. Instead of choosing between quantitative and qualitative research methods, researchers can draw on both methods to approach the subject under investigation in different ways, combine different approaches (i.e. correlational and descriptive) or triangulate, or test, findings. Take for instance a researcher who is interested in studying how the media report climate change. They could simply perform story counts to determine how many stories on climate change were published by each media outlet over a given period, and how the numbers vary between seasons and places. However, such a quantitative study might leave the researcher with a number of unanswered questions. A qualitative approach, such as closer reading of the text to determine the quality of the information in the climate stories published in the media, would add a lot of value. Additionally, researchers could carry out in-depth interviews and focus group discussions to capture the perceptions of the journalists and editors. Similarly, in the natural sciences researchers often use both quantitative and qualitative research methods either concurrently or one after the other. For instance, using quantitative research methods, researchers can determine the proportion of a given population that do not have access to a safe water source. The addition of qualitative methods, such as in-depth interviews and focus group discussions, can then help the scientists to collect more data on the perceptions and experiences of those who do not have access to safe water, and help answer explanatory questions around why access to safe water varies.

What constitutes scientific evidence?

The link between scientific research, the media and policy is illustrated in Fig 1.2. If research is conducted using a scientifically plausible methodology, the information gathered amounts to scientific evidence and it can provide a plausible basis for policy and personal decision-making. For instance, based on the evidence that second-hand smoke is harmful to non-smokers, the World Health Organization advises governments to ban smoking in workplaces and public spaces.

Naïve journalists can be deceived into reporting questionable research findings. Saunders (2013) points out key issues to consider before taking research findings as credible scientific evidence. Firstly, the research process should be objective and unbiased. Therefore, it is important for journalists to ask about the source of funding for a study when they report on research findings. Research findings may be questionable if the study is funded by an organization that stands to benefit from the research findings. If, for instance, a wine company funds a study and the findings show that a glass of wine a day is good for one's health, this should be taken with a pinch of salt. Secondly, the results should be valid and accurate. For findings to be valid, the data should be collected correctly using recognized methods. For instance, in comparing the prevalence of high blood pressure between two counties, it is important to study a similar population in both

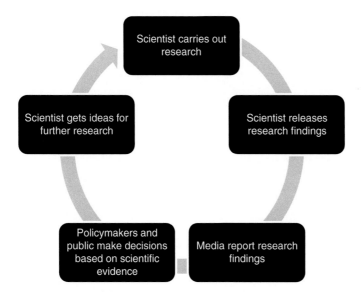

Fig. 1.2. A typical desired relationship between research and society.

counties. If the researcher examines teenagers in one county and the elderly in another county, then it is not valid to say one county has more people with high blood pressure than the other. When reporting a science story, it can be useful for a journalist to ask a scientist who is independent of the study being reported on whether the study methods and the data collected are good enough to support the conclusions.

A third crucial issue Saunders raises in regard to whether research findings should be taken as credible scientific evidence is peer review and scientific consensus. This is particularly important because science is built on the collective judgement and position of experts in a specific area of expertise, as opposed to an individual scientist's opinion. For this reason, scientists usually announce their research findings by publishing a paper in a scientific journal so that it is available for other scientists to scrutinize and use. A research paper in a scientific journal usually describes the background to the study, the study methods, the results and the conclusions. Before publishing the paper, the journal will give it to expert scientists in the same field to see whether the research was carried out using appropriate methods and whether the results provide enough evidence for the scientists' conclusion. This process is called peer review. It is important for a journalist to establish whether the findings they are reporting on have been peer reviewed and published in a reputable journal, and to specify in their story which journal published the findings and when.

One way to estimate the reputability of a journal is to look at its impact factor, which is derived from the number of times articles published in that journal have been cited by other scientists. The higher the impact factor, the more reputable a journal is. You can use online tools such as "Journal Citation Reports" to find out the impact factor of a journal. You can also seek expert scientists' views on how reputable a journal is. How long a journal has been in existence can also give you a rough idea about its reputability.

Linking scientific research and journalism

As stated above, scientists usually report their research findings by publishing articles in scholarly journals, which are mostly read by professionals in the same field. These scholarly articles usually contain technical terms that are understandable only to specialists working in that field. By accessing the scientific evidence published in journals and reporting it in the mass media, journalists provide the important service of making the information accessible to the public. This practice of reporting interesting research evidence in the mass media, in a language that is comprehensible to a non-specialist, is called science journalism. Science journalists may either report new research findings and innovations, or they may use long-established scientific knowledge to explain a current issue. Essentially, a science story focuses more on the science than on the socioeconomic and political dimensions of the issue at hand. To give an example of what makes a report a science story, consider a news report about a tornado that has struck a place for the first time in 50 years and scared onlookers. Reporting these facts alone would not qualify as a science story, but providing scientific information on why the tornado has come at this point in time, and what this implies, would.

Similarities and differences between scientists and journalists

The scientific research processes discussed in the preceding pages seem to be a whole world apart from journalism. However, upon closer scrutiny there are similarities between scientists and journalists. Colón-Ramos *et al.* (2018) point out some of these similarities. Both journalists and scientific researchers are curious; they gather information, process the information, make a conclusion based on their findings, write a report and disseminate it. However, where they differ is in their methods of collecting information, their level of precision, the time frames of their work, and their method of writing. In fact, scientists often accuse journalists of trivializing and distorting their information. Journalists, on the other hand, accuse scientists of using complicated language, taking too long to respond when asked for views or information, and being too pedantic.

Table 1.1. Differences between scientists' and journalists' processes.

Choice of words	Scientists use technical terms	Journalists use everyday words
Structure of article	Bottom-heavy. Arranged in a typical format beginning with background information and concluding with the research findings and implications. The most important information is located near the bottom of the article.	Top-heavy. The most important information, such as research findings and implications, are placed at the beginning.
Length	Most journal articles are between 5,000 and 7,000 words.	Media stories can be anywhere from one paragraph to over 2,000 words.
Level of precision	High	Moderate
Information that is prioritized	Tend to describe their methods and results in detail before giving the results and conclusions.	Most interested in the research findings and implications for society.
Mode of reporting	Report their findings by writing papers in scientific journals.	Report their findings by writing news and features articles.
Time frame	Can take many months or more than a year from start to end.	Usually take a few hours to a few days – or at most weeks – to produce a media story.

Summary

- Scientific research involves asking important questions and going through a systematic process to find and report the answers.
- The results of scientific research may be used differently by the public, professionals, policymakers and researchers.
- Scientific research can be categorized in different ways depending on the purpose of a study, the methods used and the applicability of the findings.
- The findings of scientific research should be interpreted, taking account of the context of the research methods used.
- For research findings to be considered as scientific evidence, they should be peer reviewed and published in a credible scientific journal.
- Science journalism involves reporting interesting research evidence in the mass media, in a language that is comprehensible to non-specialists.
- Although scientific research and journalism seem to be worlds apart, there are notable similarities between scientists and journalists, including their common curiosity and the fact that they both seek to publish the results of their work.

Discussion questions

1. In what ways does reporting news about scientific research findings benefit society?
2. Explain why a journalist should take note of the type of scientific research used in a study when interpreting and reporting the findings.

Suggested answers to discussion questions

Question 1

In what ways does reporting news about scientific research findings benefit society?

Suggested answers – summary of key points

- Policymakers can use the information to make better decisions for a better society.
- Individuals can use the information to make decisions to improve their lives.
- Professionals, such as doctors, engineers and agriculturalists, can use the information to improve their work.
- Researchers can use the information to get ideas for additional research.
- Industry might use the information to manufacture a new line of products.
- Teachers can use the information to update their curricula.

Question 2

Explain why a journalist should take note of the type of scientific research used in a study when interpreting and reporting the findings.

Suggested answers – summary of key points

- It helps the journalist to know whether the results have immediate application to society. The findings of basic research would most likely have no immediate application to society. Applied research, on the other hand, addresses practical problems and the findings are more likely to have immediate application to society.
- It helps the journalist to know whether they can conclude that one event causes another. Experimental research can show that one event causes another. Non-experimental research can show a relationship between two events but cannot show that one causes another. In short, correlation is not causation.
- It helps the journalist to know whether to pay more attention to numbers or perceptions in the findings. In reporting quantitative research, the story should focus on numbers, whereas in qualitative research the focus should be on experiences and perceptions.

References

Achieng, S. (2017) Africa and Asia lead in proportion of blind adults. SciDev.Net 5 October. Available at: www.scidev.net/global/news/africa-and-asia-lead-in-proportion-of-blind-adults/ (accessed 29 January 2021).
BBC (2018) More evidence essential oils "make male breasts develop". *BBC* 18 March. Available at: www.bbc.com/news/health-43429933 (accessed 29 January 2021).

Colón-Ramos, D., Kirschner, E. and Rather, D. (2018) What journalisets and scientists have in common, *Scientific American* 30 October. Available at: https://blogs.scientificamerican.com/observations/what-journalists-and-scientists-have-in-common/%20 (accessed 29 January 2021).

Frerichs, R. (2009) John Snow: British physician. *In Encyclopaedia Britannica*. Available at: www.britannica.com/biography/John-Snow-British-physician (accessed 29 January 2021).

Kishan, A.U., Dang, A., Katz, A.J., *et al.* (2019) Long-term outcomes of stereotactic body radiotherapy for low-risk and intermediate-risk prostate cancer. *JAMA Netw Open* 2(2). Available at: https://jamanetwork.com/journals/jamanetworkopen/fullarticle/2723641 (accessed 29 January 2021).

Lippia, D. and Gotuzzob, E. (2014) The greatest steps towards the discovery of Vibrio cholerae. *Clinical Microbiology and Infection* 20(3), 191–195. Available at: https://reader.elsevier.com/reader/sd/pii/S1198743X14608557?token=606E8FA380C48DE59E72B938AF56928ECB35109E3712E0E040390ECD-FF80C428F2092B8075B1C4172100EB8EDA39403A (accessed 29 January 2021).

Nyamol, O. (2020) Child malaria deaths "slashed by rainy season regimen". SciDev.Net 22 December. Available at: www.scidev.net/sub-saharan-africa/news/child-malaria-deaths-slashed-by-rainy-season-regimen/ (accessed 29 January 2021).

Saunders, M. (2013) Scientific evidence: what is it and how can we trust it? *The Conversation* 2 July. Available at: https://theconversation.com/scientific-evidence-what-is-it-and-how-can-we-trust-it-14716 (accessed 14 June 2021).

Stukeley, W. and White, A.H. (1752) *Memoirs of Sir Isaac Newton's life*. Available at: http://ttp.royalsociety.org/ttp/ttp.html?id=1807da00-909a-4abf-b9c1-0279a08e4bf2&type=book (accessed 29 January 2021).

Therrien, A. (2018) Men with low sperm counts at increased risk of illness, study suggests. BBC 18 March. Available at: www.bbc.com/news/health-43429153 (accessed 29 January 2021).

UNESCO (2015) *UNESCO Science Report: Towards 2030 launched*. UNESCO, Paris.

Vesper, I. (2017) Soil-based filter bricks clean up water for Moroccan farmers. Available at: www.scidev.net/global/news/soil-based-filter-bricks-clean-up-water-for-moroccan-farmers (accessed 29 January 2021).

World Health Organization (2019) WHO prequalifies Ebola vaccine, paving the way for its use in high-risk countries (press release). Available at: www.who.int/news/item/12-11-2019-who-prequalifies-ebola-vaccine-paving-the-way-for-its-use-in-high-risk-countries (accessed 24 December 2021).

Theories and Models of Science Communication

Dr William Tayeebwa[1], Dr Charles Wendo[2] and Dr Aisha Sembatya Nakiwala[3]

[1]*Senior Lecturer, Department of Journalism and Communication, Makerere University;* [2]*Training Coordinator, SciDev.Net, CABI;* [3]*Senior Lecturer, Department of Journalism and Communication, Makerere University*

In this chapter
- Introducing communication
- The importance of models in science communication
- Models that are specific to science communication
- Applying science communication models
- The importance of theory in science communication
- Selected theories that apply to science communication

Introducing communication

Communication is broadly understood as the process whereby an individual or a group sends a message through a channel or a medium to a receiver. There are two main goals in communication. The first is to ensure that the receiver gets the message sent by the sender without any interference. This view is problematic, however, because it takes for granted how the receiver uses the message. The second main goal of communication concerns the recipient of the message being cognitively involved in deriving meaning from the message they receive, and providing feedback as evidence that the message was understood as intended. Accordingly, communication is circular. Communication scholars believe that the success and effectiveness of the communication process is very much dependent on intervening factors that affect how the message is received, interpreted and acted upon by the recipient.

The importance of models in science communication

Communication is a complex process involving several considerations and models have therefore been developed to help illustrate, delineate and depict the structural features of communicative acts.

Scholars group communication models into four categories: transmission or linear models, ritual or expressive models, publicity or display or attention models, and reception models.

The first category, transmission or linear models, is exemplified in what is referred to as the Harold Lasswell formula, which is based on five questions: "who says what, in which channel, to whom, with what effect?" (McQuail and Windahl, 1993, p. 13). In these models, the focus is mainly put on the message sender's need to persuade. Little regard is given to the receiver of the message, since they are believed to be passive.

©2022 CAB International. Science Communication Skills for Journalists: A Resource Book for Universities in Africa (Ed. Charles Wendo)
DOI:10.1079/9781789249675.0002

The second category is the ritual or expressive model, which holds that communication is not just utilitarian but can be an end in itself. Proponents of the ritual model, notably James Carey (in 1975), have noted that "communication is linked to such terms as sharing, participation, association, fellowship and the possession of a common faith ..." (in McQuail, 2005, p. 70). In this case, communication is conceived as being aimed at uniting people in a common cause. Such a model also perceives communication as performative – for pleasure or entertainment (McQuail and Windahl, 1993, pp. 54–55).

The third category is the publicity model, sometimes known as the display or attention model. Viewed through this model, the objective of the communication process is to capture attention in order to sell a physical product or a social product (McQuail and Windahl, 1993, p. 56). In this model, the aspect of catching and holding the visual or aural attention of the audience is critical. For instance, McQuail (2005) argues that "a good deal of effort in media production is devoted to devices for gaining and keeping attention by catching the eye, arousing emotion and stimulating interest" (pp. 71–72). He further argues that this model also explains the role of the media, which is providing "diversion and passing time" (p. 72).

The fourth category is the reception model of communication propounded mainly by Stuart Hall (1980), who noted that in any communication, multiple meanings can be derived by the receiver. Hall argued that senders of messages "encode" what he called a "preferred reading", but that the receivers of the message can "decode" it in their own way and give the message a "variant or oppositional meaning" based on their experiences and outlook (McQuail, 2005, p. 73). Closely related to the above four categories are the models of science communication described below.

Models that are specific to science communication

The sets of ideas used to explain science communication are also referred to as models rather than theories. These are: the knowledge deficit model; the contextual model; the lay expertise model; and the public engagement or participation model.

The knowledge deficit model

This model, sometimes referred to as the "dissemination model", or what Secko *et al.* (2013) refer to as the "science literacy model", assumes that public scepticism about science is caused by a lack of relevant knowledge. It is based on the assumption that the experts who possess scientific knowledge are unable or unwilling to share it with the public. Several scholars, such as Brossard and Lewenstein (2010), have reviewed literature and studies that have showed a general lack of knowledge among the public regarding basic scientific concepts and facts. Science literacy programmes, including those involving media and communication, have been developed to address this knowledge gap. A key communication objective of such programmes is for scientists to develop simplified scientific information for public consumption (Brossard and Lewenstein, 2010). Secko *et al.* (2013) argue that the knowledge deficit model is pedagogically-oriented, focusing on raising science literacy levels (p. 67).

Similar to the knowledge deficit model is what Tichenor *et al.* (1970) postulated as the "knowledge gap hypothesis". After conducting several studies in the USA they noted that when information was placed in the mass media or other channels of communication, persons with more formal education acquired that information faster, and that over time the difference in the level of knowledge between the least and most educated social classes tended to increase.

The knowledge deficit model is also linked to the KAPs (knowledge, attitudes and practices) model. The KAPs model is applied in every field of inquiry to "investigate what is known, believed and done", and to "investigate experience, opinion and behaviour" (Siltrakool, 2017, p. 24). Gadzekpo *et al.* (2018) studied the understanding (knowledge), predisposition (attitudes) and response (practices) of Ghanaian journalists in relation to reporting on climate change. The researchers established that there was "high awareness, but low knowledge, high conviction but low engagement of media practitioners and institutions towards addressing the incidence and consequences of climate change" (p. 1). They further noted that while awareness and empathy were prerequisite factors for reporting on climate change, they "did not inevitably lead to knowledge and engagement in competently mediating the climate message" (p. 1).

David Dickson (2005), founder and former editor of SciDev.Net, concluded in a commentary about the knowledge deficit model as follows:

> Increased knowledge about modern science does not necessarily lead to greater enthusiasm for science-based technologies. Indeed, there is considerable evidence to the contrary. For example, the more knowledge an individual has about a potentially dangerous technology (such as nuclear power or genetic engineering), the more concern he or she may well feel about that technology.

While the knowledge deficit model has several discernible strengths, as highlighted above, its main shortcoming is its inclination to view communication as one way. Further, perceptions of, and the utilization of, scientific information are more complex than is portrayed in the deficit model. The model also ignores the sociocultural and material contexts that play a major role in the understanding of scientific information.

The contextual model

This model is also called the public engagement model. It proposes that effective science communication requires an understanding of the needs, attitudes and existing knowledge of different audiences (Lewenstein, 2003). Brossard and Lewenstein (2010) point out that the contextual model "acknowledges that individuals do not simply respond as empty containers to information, but rather process information according to social and psychological schemas that have been shaped by their previous experiences, cultural context, and personal circumstances" (pp. 13–14). The context in which individuals receive information on science, and the social forces in society, is also important (Lewenstein 2003, p. 3). Media systems and other societal or institutional circumstances are intervening factors in regard to how messages about science are transmitted. However, like the deficit model, the contextual model also tends to accord too much power to scientists, thus viewing the communication process one way, with hardly any interaction between the source of the scientific information and the recipients (Brossard and Lewenstein, 2010, p. 14).

The lay expertise model

This model calls for an appreciation of local expertise and knowledge regarding scientific subjects under consideration. It foregrounds the significance of the tacit knowledge possessed by communities through, for example, elders and other opinion leaders. Brossard and Lewenstein (2010) argue that there is vast knowledge based in the "lives and histories of real communities" (p. 15). They argue that "communication needs to be structured in ways that acknowledge information, knowledge and expertise already held by communities facing scientific and technical issues" (p. 14). Lewenstein (2003) argues that the "lay-expertise model" puts an emphasis on "knowledge and expertise that is held and validated by social systems other than modern science" (p. 5). However, Brossard and Lewenstein (2010) argue that the lay expertise model aims at placing value on local knowledge as genuine expertise in its own right (p. 15). This they argue is unlike conventional approaches to indigenous knowledge systems, which often use modern science methods to verify traditional beliefs. One of the key criticisms of the lay expertise model is that it devalues the scientific approach whereby knowledge is produced through empirical inquiry, in favour of "lay knowledge" that is often based on hunches and unproven claims.

The public engagement or participation model

In this model of science communication, the scientists, the public and policymakers participate equally in discussions and debates about issues in science and technology. The model is applicable to situations where collaboration is the desired communication objective. As conceptualized in this model, science communication may take the form of participatory engagements to ensure audiences are central in shaping the future of research efforts. Such public participation activities could include community conferences, town hall meetings, community workshops and symposia. Brossard and Lewenstein (2010) point out that "public participation activities are often driven by a commitment to 'democratizing' science – taking control of science from elite scientists and politicians and giving it to public groups through some form of empowerment and political engagement" (p. 16). As in the case of the lay expertise model, the public engagement model has been criticized for diminishing and sometimes downplaying the expertise of scientists (Lewenstein, 2003, p. 6).

Table 2.1 summarizes the basic elements and weaknesses of each of the four models of science communication outlined above.

Applying science communication models

Some scholars – for example, Cormick (2019) – argue that the deficit model should not be used because it is erroneous. Cormick's argument is that people are not empty vessels waiting to be filled with information. Rather, they have their own ideas, beliefs and knowledge that influence how they receive and perceive new scientific information. Effective science communication is not a matter of translating scientific knowledge into simpler information that the public can easily understand. Cormick argues that the deficit model should be "buried" (p. 12).

However, other scholars, such as Hetland (2014), have argued that models of science communication are not mutually exclusive but rather are interrelated. Indeed, in real life, the dissemination of information is usually the starting point for consultation and engagement.

The importance of theory in science communication

In lay person's language the word theory can be used to refer to a guess or an assumption. In academia, on the other hand, theory refers to a tested set of concepts, explanations and principles that make up the body of knowledge in a field of study, such as mass communication (Baran and Davis, 2012, p. 24). While there are many academic definitions of the term theory, the common element among the various definitions is that theories explain phenomena or predict outcomes by describing the relations between variables. Shoemaker *et al.* (2004, p. 112) argue that when we move from description or modelling to understanding how the object or process or system works, we move to theory.

As Baran and Davis (2012) explain, theories of communication vary depending on how one wants to understand the various functions or elements of the communication process. For instance, each element of the communication process – from the sender, to the message, to the channel or media, to the recipient – necessitates specific inquiry and understanding. For example, Stuart Hall's reception theory predicts that different people will perceive a single message in different ways, and their perception may not necessarily match what was meant by the sender.

Table 2.1. Basic elements and weaknesses of models of science communication (adapted from Brossard and Lewenstein 2010, p. 17; Secko *et al.* 2013, p. 67).

Model	Basic elements	Weaknesses
1. Knowledge deficit model	• Scientists are experts and are knowledge-able. • The public have a deficiency of knowledge. Delivery of simplified scientific information leads to public understanding and acceptance of science. • Transfer of knowledge is one way, from scientists to the public. • Good transmission of scientific information leads to a reduced deficit in knowledge. • A reduced knowledge deficit leads to better decisions, and often better support for science.	• Perception and utilization of scientific information is more complex than portrayed in the deficit model. • Overlooks importance of background knowledge and sociocultural circumstances in science communication. • The public is not homogeneous. Reception of information will vary from person to person.
2. Contextual model	• Communication of science is considered to be based on the needs, attitudes and existing knowledge and situations of the different audiences. • Individuals respond to messages based on their unique circumstances. • There is one-way transmission of information from scientists to the public. • Audiences have ability to quickly gain knowledge about topics that are relevant to them.	• According to this model, communication is one way: no interaction between the source and recipient of knowledge. • Absence of adequate opportunity for feedback.
3. Lay expertise model	• Acknowledges the limitations of scientific information. • Acknowledges that audiences might have some pre-existing knowledge. • Highlights interactive nature of scientific process.	• Undermines the expertise of scientists.
4. Public engagement or participation model	• Two-way flow of information between scientists, the public and policymakers. • Communication strengthens relations between science and the public. • Focuses on policy issues involving scientific and technical knowledge. • Tied to democratic ideal of wide public participation in policy process. • Builds mechanisms for engaging citizens in active policymaking. • Real public authority over policy and resources.	• Diminishes the scientist's power. • Citizens can participate in a more emotional than rational way, which can undermine the objective of communication. • More complex, and therefore difficult to explain to donors and policymakers.

Perry (2002) argues that a good theory is developed through scientific inquiry or research, and provides several functions, including aiding understanding of the causes of events, enabling predictions of the future, and providing the foundation for further scientific knowledge, among others (p. 43). Theories of communication can help in predicting how people are likely to receive, perceive and respond to information about science. Such theories also help support the appreciation of recommendations regarding the basic principles of effective science communication.

Selected theories that apply to science communication

In this section we present four theories we consider demonstrative of the relationship between scientists and communication/journalism institutions.

Reception theory

The key proponent of reception theory was Stuart Hall, who in his reception studies noted how audiences made varied interpretations of specific forms of content (Baran and Davis, 2012, p. 304). He noted that while reading "encoded" texts, audiences often interpret or "decode" them in different ways. While the producers of the messages usually have a "preferred or dominant reading", sometimes audiences have alternative interpretations of the messages, in what Hall referred to as a "negotiated meaning", which often differs from the intended meaning. This process has been referred to as the encoding-decoding of discourse (McQuail, 2005, p. 117).

Baran and Davis (2012) highlight several strengths of reception theory, including the fact that it focuses on individuals in the mass communication process and that it respects the intellect and ability of media consumers (p. 305). Further, the theory "acknowledges a range of meanings in media texts" and "seeks an in-depth understanding of how people interpret media content" (p. 305). They also argue that the theory "can provide an insightful analysis of the way media are used in everyday social contexts" (p. 305). However, according to Baran and Davis the theory has some weaknesses, including the fact that "it is usually based on subjective interpretation of audience reports" and "cannot address presence or absence of effects" (p. 305).

For science journalists and communication professionals, reception theory draws attention to the fact that messages can be interpreted in ways that were not initially intended. It highlights the agency of the receivers of the messages and the importance of paying attention to their characteristics to ensure a better encoding of messages. It also brings to light the importance of pre-testing messages.

Agenda-setting theory

Earlier communication scholars, such as Lippman, in 1922 (Baran and Davis, 2012), pointed out the power the media hold in regard to influencing public opinion. Later propaganda scholars, such as Herman and Chomsky (1988), confirmed how powerful elites are able to use the media to propagate their limited concept of reality to the majority. The agenda-setting theory advances the idea that "media do not tell people what to think, but what to think about" (Baran and Davis, 2012, p. 346). Cohen (1963), one of the first proponents of the agenda-setting theory, presented the theory in a foundation statement as follows:

> The press is significantly more than a purveyor of information and opinion. It may not be successful much of the time in telling people what to think, but is stunningly successful in telling its readers what to think about. And it follows from this that the world looks different to different people, depending not only on their personal interests, but also on the map that is drawn for them by the writers, editors, and publishers of the papers they read.
>
> (Cohen, 1963, p. 13, in Baran and Davis, 2012, p. 347)

Other scholars, notably Maxwell McCombs and Donald Shaw, conducted research to test the theory and came to the following conclusion, in yet another foundation statement:

> In choosing and displaying news, editors, newsroom staff, and broadcasters play an important part in shaping political reality. Readers learn not only about a given issue, but how much importance to attach to that issue from the amount of information in a news story and its position … The mass media may well determine the important issues – that is, the media may set the 'agenda'
>
> (Baran and Davis, 2012, p. 347)

In the context of science journalism and communication, it follows that the scientific issues and topics that the media decide to focus on inevitably become part of the public's agenda. The agenda-setting theory emphasizes the power of the media in guiding people on what to think about, but also recognizes the agency of the public in deciding for themselves what to think. For science journalists and communication professionals, this power, highlighted in the theory, can be harnessed to direct public discourse towards critical issues, such as climate change, public health during global pandemics such as the Coronavirus pandemic, and other topics.

Framing theory

Framing as a theoretical framework was expounded by Entman (1993), who explained it as follows: "to select some aspects of a perceived reality and make them more salient in a communication text, in such a way as to promote a particular problem definition, causal interpretation, moral evaluation, and/or treatment recommendation" (p. 52). The key features in this conceptualization are not only the selection process, but also – and even more so – the intention of the frame sponsors, which is to influence at different levels how an issue is considered for action. Valkenburg *et al.* (1999) define a frame as "a schema of interpretations that enables individuals to: perceive, organize, and make sense of incoming information" (p. 551).

Science journalists and communication professionals can take from framing theory the idea that how topics and issues are portrayed to the public matters a great deal. For example, if the most prevalent voice is that of vaccine pessimists who present a vaccine as a danger to human health rather than a much-needed solution to a health crisis, then the chances are that resistance to receiving such vaccines will be greater.

Diffusion of innovations theory

Baran and Davis (2012) argue that the diffusion of innovations theory (Rogers, 1995) can be seen as an extension of Paul Lazarsfeld's two-step flow theory. Rogers used several studies to show that scientific innovations go through several stages before being widely adopted. They summarize the stages as follows (p. 332):

> First, people become aware of them, often from mass media information. Second, the innovations get adopted by a small group of innovators or early adopters. Third, opinion leaders learn from the early adopters and try the innovation themselves. Fourth, if opinion leaders find them useful, they encourage their friends – the opinion followers. Finally, after most have adopted the innovation, a group of laggards, or late adopters, makes the change.

In a later reformulation of the theory to explain the information flow of the diffusion, McQuail (2005, p. 490) pointed out that Rogers and Shoemaker (1973) postulated four stages: knowledge (whereby the individual is exposed to an awareness of the existence of the innovation so as to gain understanding of how it functions); persuasion (the individual forms a favourable or unfavourable attitude towards the innovation); decision (the individual engages in activities which lead to a choice to adopt or reject the innovation); and confirmation (the individual seeks reinforcement for the innovation decision he or she has made, but may reverse the previous decision if exposed to conflicting messages about the innovation) (McQuail and Windahl, 1993, p. 74).

Arguably, the diffusion of innovation theory is one of the most applicable theories for science communication and journalism, and is closely related to the knowledge deficit model of science communication. While the theory has been widely criticized for its top-down and linear nature, the communication of science and technology remains deeply entrenched in the diffusion tradition.

Conclusion

There has been significant debate within the scholarly community on the application of the different theories and models of science communication. Some scholars, such as Lewenstein (2003), have suggested that these models provide a framework for understanding how the public can engage better with science, and how scientists can make their work more accessible (p. 7). We have attempted to show that models in which science communication is conceived as a linear process are problematic because they do not consider intervening factors that influence how the public interpret the information they receive about science. By paying attention to the recipients of science information, as well as their sociocultural and material contexts, it is possible to communicate in ways that make information about science more understandable and acceptable. In this connection, Faehnrich (2021) appraises the changing media landscape and the impact the digital transformation is having on science communication. He notes the advantages the new channels of communication – such as online social networking sites – offer for science communication, including increased space for public engagement and the diversification of actors who engage in science communication (p. 2). Secko *et al.* (2013) apply the four models of science journalism and communication to journalistic coverage of science and propose several criteria for such application (p. 73). Recently, Khairy (2020) has used the four models of science journalism to understand how the public interacted with news about the Coronavirus pandemic. Thus, from our discussion it is evident that models and theories of science communication and journalism remain relevant not only for scholars but also for practitioners as well.

Summary

- Theory helps predict how people are likely to receive, perceive and respond to information about science.
- Whenever we communicate, we are trying to achieve one or more of the following: to inform, persuade, consult or engage. Theories and models of science communication are aligned to these purposes of communication.
- Several normative theories of communication are applicable to science communication as well: for example, diffusion of innovation theory, reception theory, the knowledge gap hypothesis, framing theory and agenda-setting theory.
- There are four key models that are specific to science communication: the knowledge deficit model, the contextual model, the lay expertise model and the participation model. These differ in regard to the level of audience involvement and consideration of sociocultural factors that influence the way people perceive information. However, some scholars argue that these models of science communication are interrelated and not mutually exclusive.

Discussion questions

1. Discuss the strengths and weaknesses of each of the four models of science communication.
2. Explain the importance of theory in science communication.

Suggested answers to discussion questions

Question 1

Discuss the strengths and weaknesses of each of the four models of science communication.

Suggested answers – summary of key points

Knowledge deficit model

Strengths: A good model for quickly delivering information to large numbers of people and causing mass awareness. Takes into account scientists' expert knowledge.

Weaknesses: Awareness alone may not lead to action. Sees communication as one way. Does not take into account how people's perceptions depend on their sociocultural backgrounds and prior knowledge.

Contextual model

Strengths: Takes into account how people's needs, attitudes, existing knowledge and situations affect their responses.

Weaknesses: The emphasis is still on one-way communication; audience feedback is ignored.

Lay expertise model

Strengths: Takes into account the background knowledge and expertise of the recipients.

Weaknesses: It might undermine the expertise of scientists.

Participation model

Strengths: Takes into account knowledge, experiences and feedback of the target group.

Weaknesses: Participation can be more emotional than evidence-based discussions.

Question 2

Explain the importance of theory in science communication.

Suggested answers – summary of key points

- Theory helps predict how people are likely to receive, perceive and respond to information about science.
- Theory helps in understanding and explaining the outcomes of specific communication interventions.
- Theory helps in designing and planning effective science communication.
- Theory provides the foundation for further research into science communication.

References

Baran, S.J. and Davis, D.K. (2012) *Mass Communication Theory: Foundations, Ferment and Future*, 6th edn. Wadsworth Cengage Learning, Toronto, Canada.

Brossard, D. and Lewenstein, B.V. (2010) *A Critical Appraisal of Models of Public Understanding of Science: Using Practice to Inform Theory*. Routledge, London, United Kingdom.

Cohen (1963), in Baran, S.J. and Davis, D.K. (2012) *Mass Communication Theory: Foundations, Ferment and Future*, 6th edn. Wadsworth Cengage Learning, Toronto, Canada.

Cormick, C (2019) *The Science of Communicating Science: The Ultimate Guide*. CABI, Wallingford, United Kingdom.

Dickson, D. (2005) The case for a "deficit model" of science communication. SciDev.Net 24 June. Available at: https://perma.cc/PG4T-3V6S (accessed 22 June 2021).

Entman, R. M. (1993) Framing: Toward clarification of a fractured paradigm. *Journal of Communication* 43, 51–58.

Faehnrich, B. (2021) Conceptualising science communication in flux – a framework for analysing science communication in a digital media environment. *Journal of Science Communication* 20(03). Available at: https://perma.cc/8WGU-Z7BN (accessed 24 December 2021).

Gadzekpo, A., Tietaah, G.. Segtub, M. and Segtub, M. (2018) Mediating the climate change message: knowledge, attitudes and practices (KAP) of media practitioners in Ghana. *African Journalism Studies* 39(3), 1–23.

Hall, S. (1980) Encoding/decoding. In: Hall, S., Hobson, D. and Willis, P. (eds). *Culture, Media, Language*. Hutchinson, London, United Kingdom, pp. 128–138.

Herman, E.S. and Chomsky, N. (1988) *Manufacturing Consent: The Political Economy of the Mass Media*. Pantheon, New York, United States.

Hetland, P. (2014) Models in science communication policy. *Nordic Journal of Science and Technology Studies* 2(2), 4–17.

Khairy, L. (2020) Applying the four models of science journalism to the public's interaction with Coronavirus news. *Arab Media and Society*. Available at: https://perma.cc/257U-F6HQ (accessed 24 December 2021).

Lewenstein, B.V. (2003) Models of public communication of science and technology. *Public Understanding of Science*. Available at: https://perma.cc/S8ZL-TPKB (accessed 24 December 2021).

McQuail, D. (2005) *McQuail's Mass Communication Theory*, 5th edn. Sage Publications, London and New York.

McQuail, D. and Windahl, S. (1993) *Communication Models for the Study of Mass Communication*, 2nd edn. Pearson Prentice Hall, London, United Kingdom.

Perry K.D. (2002) *Theory and Research in Mass Communication: Contexts and Consequences*, 2nd edn. Lawrence Erlbaum Associates, Inc, New Jersey, USA.

Rogers, E.M. (1995) *Diffusion of Innovations*, 4th edn. The Free Press, London, United Kingdom.

Rogers, E.M. and Shoemaker, F. (1973) *Communication of Innovations*. The Free Press, Glencoe, United States.

Secko, D.M., Amend, E. and Friday, T. (2013) Four models of science journalism. *Journalism Practice*, 7(1), 62–80.

Shoemaker, P.J., Tankard, J.W. and Lasorsa, D.L. (2004) *How to Build Social Science Theories*. Sage Publication, London, United Kingdom.

Siltrakool, B. (2017) Assessment of community pharmacists' knowledge, attitude and practice regarding non-prescription antimicrobial use and resistance in Thailand. Available at: www.researchgate.net/publication/322675323 (accessed 24 December 2021).

Tichenor, P.J., Donohue, G.A. and Olien, C.N. (1970) Mass media flow and differential growth in knowledge. *Public Opinion Quarterly* 34, 159–170.

Valkenburg, P.M., Semetko, H.A. and Vreese, C.H.D. (1999) The effects of news frames on readers' thoughts and recall. *Communication Research* 26, 550–569.

Current Status and Future of Science Journalism and Communication in Africa

Dr Abraham Kiprop Mulwo
Associate Dean, School of Information Sciences and Senior Lecturer Department of Publishing, Journalism and Communication, Moi University, Eldoret, Kenya

In this chapter
- The genesis of science and science journalism
- Current trends in science journalism in Africa
- Science journalist associations
- Science desks in the media
- Science-related programming in the media
- Science journalists
- Challenges faced by science journalists in Africa
- Preparing for the future of science journalism in Africa

The genesis of science and science journalism

Science and technology are embedded in the everyday experiences of contemporary living. As such, societies are increasingly finding it important to incorporate scientific facts in decision-making processes on topical issues such as nutrition, healthy living and climate change, among others. However, effective communication of science is a complex endeavour that requires skill. In essence, it requires specific adaptations of communication to specific audiences with differing circumstances (National Academy of Sciences, 2017).

It has been argued that science communication emerged in the 17th century, when the idea of taking evidence as the basis for knowledge first became popular. Before the advent of science, society believed in religious authority and the wisdom of philosophers such as Aristotle.

Galileo Galilei's dialogues of 1632 and 1638 demonstrate how a lack of data, methodology and scientific language undermined his efforts to communicate his ideas (Arianrhod, 2019). In the second phase of the 17th century, however, science and scientific writing became more popular, even though the technology and methods of scientific inquiry at the time were still quite primitive. This later gave way to experimentations and advancements in philosophy and science (Artal, 2015).

The history of science communication can be traced back to Ancient Greece, when knowledge was imparted in public places and was deliberated on by the masses. As science became increasingly recognized as the basis for decision making, from the 17th century onwards, the need for science communication increased. However, scientists faced difficulties in turning complex scientific data into information that non-specialists could understand and relate to. For instance, while Isaac Newton's publication *The Principia* provided empirical and theoretical facts that could be tested, its extensively mathematical language could only be understood by

©2022 CAB International. Science Communication Skills for Journalists: A Resource Book for Universities in Africa (Ed. Charles Wendo)
DOI:10.1079/9781789249675.0003

mathematicians of Newton's mettle, thus undermining his efforts to communicate his scientific findings. With the advancement of scientific methods and communication technologies, scientific communication continued to be elitist, making it difficult for ordinary citizens to access scientific information (Raza, 2013). Furthermore, hand-copied manuscripts and books were so expensive that citizens could not afford to buy them, even if they could read them.

Modern science communication emerged out of the growing need to make scientific information accessible to ordinary citizens. Even though it is now an established field, its rate of growth has varied across the world. In Australia, for instance, science communication has evolved dramatically in the past 60 years, giving rise to interactive science centres, university-level courses that teach the theory and practice of scientific communication, and a rapid increase in the employment of scientific communicators. In this chapter, we discuss the current status of science communication in Africa, with a special focus on science journalism.

Current trends in science journalism in Africa

In Africa, as elsewhere across the globe, media reporting is overwhelmingly dominated by political, sports and business news. Stories that focus solely on science rarely make news.

Nevertheless, science-related disciplines such as health, environment, agriculture and energy have in recent years attracted growing interest among African media institutions. This has been spurred by major concerns such as climate change, epidemics and non-communicable diseases, which continue to have a huge impact on communities across the world. For example, the global impacts of climate change continue to unfold and information on this subject is therefore of interest to audiences globally. Similarly, current health challenges, such as the COVID-19 pandemic, continue to dominate the media landscape and highlight the importance of science journalism and communication.

Nakkazi (2012) reports on the growth of science journalism in Africa from the early 2000s. Whereas science journalism in the Global North was experiencing a crisis during this period, with science desks shutting down and science journalists changing to other news beats, the reverse was true in Africa. Editors in African countries cited an improvement in the number of journalists reporting science stories, the quality of stories and the number of media outlets with dedicated science space. Nakkazi attributes the growth of science journalism in Africa to the activities of professional associations: for example, SjCOOP, a science journalism training and mentoring programme run by the World Federation of Science Journalists. About 100 African journalists benefited from the programme between 2006 and 2012. During the same period several new science journalism associations were formed in Africa, and scientists' trust in journalism increased. Lugalambi *et al.* (2011) also reported an improvement in the trust and engagement between scientists and journalists over time, with scientists being more willing than before to share information with journalists.

This positive outlook of science journalism in Africa was confirmed by the *Global Science Journalism Report* (Massarani *et al.*, 2021). According to the report, science journalists in Africa were more satisfied with their work than those in most other parts of the world, even though most of them worked as freelancers, as opposed to being staff reporters.

The media plays a critical role in shaping public opinion on topical issues, including scientific developments. For instance, the way a society perceives a health crisis is, to a large extent, determined by how the media reports that crisis. Research has shown that the media is the main source of information for most people. However, despite the improvements cited above, in

Africa, as elsewhere across the world, the coverage of scientific information in the media is still disproportionately low when compared to other issues, such as politics, sports and business. It has also been noted that, of the few science stories that do get covered in the media, many often lack contextual information on the scientific issue being reported, while others quote press releases verbatim, exhibiting a lack of analytical reporting of the issue at hand. Several reasons account for the low reporting of science in the mainstream media in Africa, including the following.

- *The perceived unpopularity of science stories* – Most science journalists complain of the lack of interest in science stories among editorial teams in newsrooms. Editors give less space to science stories than to politics and other newsbeats.
- *Competition for space in the media* – Science news competes for space with other stories. The general perception among some media practitioners is that the public have limited interest in receiving scientific information because they would not understand it anyway.
- *Skills deficit* – Most journalists do not have adequate skills to report accurate, comprehensible and compelling science stories. Few journalists have a science background, while training institutions across Africa have not taken measures to adequately equip journalists to report on science matters. Similarly, scientists rarely have knowledge about how the media operates, and are therefore unable to repackage their scientific reports in a manner that will make it easy for the media to review them and publish stories on them.
- *Scientific jargon* – Useful research findings published in scientific journals or presented at scientific meetings are often presented in scientific jargon, which journalists do not find easy to understand.
- *Access to scientific journals* – Most scientific journals are only accessible via subscription and thus journalists and media houses often do not have access to the research published in such journals.

Despite this negative picture, with the increase in the number of journalists that have undergone training on science reporting, coupled with the trend of fast-evolving news-making stories, especially in the areas of public health and climate change, science stories are increasingly receiving wider coverage in many African countries. A survey conducted by the African Academy of Sciences noted that the coverage of science stories in Kenya, Senegal, South Africa and Nigeria has rapidly increased since 2014 (Waithera, 2018). The study further established that climate change and food and nutrition constitute the bulk of science stories covered in these four African countries. On the other hand, science, technology, engineering and mathematics (STEM), sustainable energy, health and wellbeing, among others, receive little coverage. Most science stories appear in the news segment, while a significant proportion feature in opinion/editorial sections.

The *Global Science Journalism Report* (Massarani *et al.*, 2021) found that 89% of science journalists in Africa have a university degree, with most (57%) having a degree in journalism, as opposed to sciences (13%). This is different from Europe, the USA, Canada, and Asia and the Pacific, where most science journalists have a degree in sciences. The report found that most science journalists in Africa have six to ten years of experience – in contrast to Europe, the USA, Canada, Latin America, and Asia and the Pacific, where most have more than 15 years' experience. Regarding working conditions, the *Global Science Journalism Report* found that more science journalists in Africa work as full-time freelancers than as full-time staff. The respondents reported that their workload had increased but their working conditions were better than before. In a nutshell, the report paints a promising picture of science journalism in Africa, with younger and busier – and also happier – science journalists compared to other continents.

Science journalist associations

A number of science journalist associations are emerging in Africa, signalling renewed interest in the discipline. Table 3.1 shows some of the science journalist associations in sub-Saharan Africa. These associations help journalists to share ideas, support each other and create learning opportunities with a view to enhancing the coverage of science stories in the media.

Science desks in the media

A science desk refers to a team of journalists within a media outlet whose primary focus is to report science stories. Science desks are the engine of science stories for media outlets. The existence of a science desk indicates a media organization's commitment to reporting on science and technology. It also helps the media organization to build a team of journalists with the competence to report science. However, such desks are known by different names in different media houses, depending on how the media house is structured. For instance, at Standard Media in Kenya, the science desk is housed under the national news section, with one editor managing both health and gender reporting. Similarly, the Nation Media Group has a science desk that is dedicated to integrating the coverage of science and health for print, broadcast and digital media platforms.

Nevertheless, concerns have been raised over the lack of dedicated science desks among smaller media houses, including local language FM radio stations and television stations, which have wide coverage at the grassroots level. Such media outlets play a key role in shaping public discourse. A key opportunity to influence public opinion on science matters is missed due to the lack of dedicated science desks within these outlets.

Science-related programming in the media

Some broadcast media outlets have programmes that are geared towards empowering their audiences by providing them with scientific information. It is not clear what proportion of media

Table 3.1. Some networks of science journalists in sub-Saharan Africa.

Media association	Acronym	Country
Association des Journalistes Scientifiques de Guinée	AJSG	Guinea
Association des Journalistes Scientifiques du Niger	AJSN	Niger
Association des Journalistes Scientifiques et Communicateurs pour la Promotion de la Santé	AJC-PROSANTE	Cameroon
Kenya Environment and Science Journalists Association	KENSJA	Kenya
Media for Environment, Science, Health and Agriculture	MESHA	Kenya
Medical Journalists' Association Ghana	MJAG	Ghana
Nigeria Association of Science Journalists	NASJ	Nigeria
Rwanda Association of Science Journalists	RASJ	Rwanda
Science Journalists and Communicators of Togo	JCS-Togo	Togo
Somali Media for Environment, Science, Health and Agriculture	SOMESHA	Somalia
Uganda Science Journalists Association	USJA	Uganda
Zimbabwe Environmental Journalists Association	ZEJA	Zimbabwe

outlets in the region have such programmes; however, where they exist, they tend to be popular with audiences. For instance, in mid-2021, Inooro television, a vernacular television station owned by Royal Media Services in Kenya, ran a successful interactive television programme called *Ndagitari*. In the programme, medical professionals were invited to the studio to lead a conversation on topical health issues. Viewers were invited to phone in and participate in the discussion. The programme ran for one hour every weekday and was described as one of the most popular vernacular television programmes on air. Similarly, Inooro radio ran two successful programmes called *Mugambo wa Murimi* (*Voice of the Farmer*) and *Uthunduri* (*Science and Innovation Programme*) for ten minutes every weekday. The popularity of these programmes demonstrates that, when they are properly structured so as to demonstrate relevance to target audiences, science programmes have the potential to attract a large market share among audiences.

Science journalists

The number of African journalists who report about science has increased in recent years, partly due to the activities of science journalism associations, as pointed out by Nakkazi (2012). Several universities have also recently revised their curricula to include science journalism courses, such as health reporting, environmental journalism, science journalism and so on. In addition, specialist courses, such as Science Communication Skills for Journalists, run by Script in collaboration with Makerere University (Uganda), Nasarawa State University Keffi (Nigeria), Moi University (Kenya) and University of Dar es Salaam (Tanzania), have contributed to addressing capacity gaps in science journalism by offering additional training on science reporting to journalism students in those universities.

Media institutions have also enhanced their science journalism capacities by hiring reporters with a science background. At Standard Media in Kenya, as at mid-2021, Dr Mercy Korir, a trained medical doctor who is now pursuing a career in the field of journalism, was in charge of Health and Science for the outlet. Journalists like Dr Korir have contributed immensely to enhancing the media's capacity to report on science issues. Dr Korir has been key in helping media institutions in Kenya to report accurately on COVID-19 and to debunk the myths and fake news surrounding the disease.

However, skilled science reporters remain few in number in Africa and some media houses are forced to rely on reporters who have developed an interest in covering science stories but who are not trained in this area. Such untrained journalists have sought to enhance their capacities and to advocate for the broader inclusion of sciences in media coverage by joining science journalist associations (see the earlier section on such associations).

Challenges faced by science journalists in Africa

While science reporting has grown in the past few years, African journalists who report science stories still face several challenges, including the following.

- *Lack of training opportunities* – As indicated above, most journalists who currently report on science stories have had little or no formal training on science reporting, owing to the lack of relevant courses at media training institutions across the continent. To bridge the gap, a number of them have undertaken short courses organized by their media houses and/or science journalist associations.
- *Heavy workload* – Some science journalists have reported burnout as a result of having to cover other newsbeats in addition to science. This is because most media houses are either

unwilling or unable to hire dedicated science journalists. As a result, interested journalists pursue the non-science stories assigned to them by their editors and then report on science subjects during their spare time. When the science stories get rejected after such extra efforts, it adds to their frustration.

- *Declining interest in the news media* – Reports have shown that a general decline is taking place in audiences' levels of trust in news across the world. This, coupled with rapidly growing alternative sources of information, leads to reduced interest in the news media. In turn, this affects the revenue of media organizations.
- *High turnover of science journalists* – African media continues to report a high turnover of science journalists, who either change to other newsbeats or quit the media sector to pursue other careers. Many of these journalists report frustrations relating to the challenges of getting their stories published.
- *Access to sources of information* – Science journalists seeking to obtain information face a daunting task, owing to the scientific jargon used in science journals. This is further complicated by difficulty in accessing cutting-edge reports in established journals, which are not open access. In addition, most scientists have limited interest in media coverage, since they fear that journalists might misrepresent their research (mostly as a result of editing processes in newsrooms). To mitigate this challenge, some research institutions have now hired information officers who assist in preparing reports and press releases for the media.
- *Lack of interest in science stories in newsrooms* – One of the key challenges often cited by science journalists is the lack of interest in science reporting among editorial teams, who are the main gatekeepers in media houses. Many science journalists complain that they spend several hours developing science stories only for them to be discarded or given a brief slot towards the end of a news broadcast. Journalists can overcome this by developing skills in pitching science stories in such a way that the editor can clearly see the relevance and urgency.
- *Competition with other newsbeats* – Science reporting faces stiff competition from other newsbeats, such as politics, sports, business and entertainment. Owing to the limited understanding of science journalism among many editorial teams, science stories often receive limited consideration.

The most common reasons cited for the low quantity of science reporting in the news media in Africa are summarized in Box 3.1.

Preparing for the future of science journalism in Africa

Firstly, there is a need to advance science journalism as a discipline within the communication sector, with a view to enhancing the skills of both scientists and journalists who report on scientific information (Odlyzko, 2000). Faced with global challenges such as climate change, disease and

---

Box 3.1. Reasons cited for low quantity of science reporting in African media.

- Most journalists do not understand scientific jargon
- Most journalists cannot access scientific journals
- Most journalists do not have adequate skills to report science stories
- Science is out-competed by other more sensational topics
- Low prioritization by editors

economic shifts, among others, the recognition of science reporting as a fundamental pillar in policymaking is gaining currency. A growing number of universities in the western world are now offering science communication as a degree programme. In Africa, a number of universities have integrated science journalism into their training of journalism students (Table 3.2).

Secondly, there is also a need for scientists and journalists to have a mutual understanding of how each conceptualizes scientific work (Berenbaum, 2017). A journalist is concerned with a story. For instance, what is new and why should the public care about it? What is it that will excite the reader/audience? On the other hand, scientists are often driven by their desire to answer their – often complex – research questions. The challenge therefore is to train scientists to be able to simplify their research findings so that they become understandable and interesting to the reader, and therefore the journalist. Myers and Worm's (2003) publication in *Nature* titled "Rapid world-wide depletion of predatory fish communities" is a good example. In this article, Myers and Worm demonstrate, in a simplified way, how industrialized fisheries reduce the population of the large predatory fish biomass by 80% within 15 years of exploitation, and they highlight the consequences of this depletion for the marine ecosystem. This is an appealing story for the media and, needless to say, it attracted the attention of major media outlets across the world. Scientists and journalists alike need to be taught how to make a complex story simple.

Thirdly, leveraging new media technology, there is a need for science–media engagement at all levels of research, not just in the final phase where research findings are disseminated (Jothi and Neelamalar, 2010). In the current new media environment, there are several tools and platforms through which scientists can engage with the media and the general public at all stages of their research work. Social media platforms, for instance, can help scientists draw the attention of the media, policymakers and the public to the kind of research they are currently engaged in, and to build interest in it by demonstrating its relevance to society (Zhu and Purdam, 2017).

Fourthly, there is a need to encourage policy-engaged research in African institutions. Science and policy are intertwined: science attempts to understand problems while policy aims to solve problems. Research attracts the attention of the media, policymakers and the general society when researchers can demonstrate its key policy relevance. Policymaking is a murky process, but evidence is key in this process. Science serves to provide this evidence. Thus, scientists need to understand policymaking processes and be able to identify specific windows of opportunity when they can use scientific findings to influence policy.

Fifthly, there is a need to encourage greater exchange of knowledge on science and science reporting through organizing regular conferences. Scientific conferences provide platforms for scientists to share their research findings. By participating in such conferences, science journalists get the opportunity to enhance the coverage of their scientific research, ensuring that their research is used for the benefit of humanity.

Finally, previous research looking at science journalism has concentrated mostly on the capacity of journalists to report on scientific findings, ignoring the gatekeeping role of media editors

Table 3.2. African universities that have integrated science journalism and communication into their training.

University	Country
Nasarawa State University, Keffi	Nigeria
Makerere University	Uganda
Moi University	Kenya
University of Dar es Salaam	Tanzania

and media owners. Media reporting is, to a great extent, determined by the business interests of media owners, who, together with media editors, set the agenda for media coverage. There is therefore a need to focus on media editors and media owners as key players in determining the reporting of scientific findings in African media.

Summary

- Science communication existed as early as the 17th century or before, and has since evolved, following the developments in media technology.
- While the coverage of scientific findings in the news media has declined globally in recent years, recent studies show that there has been a renewed interest in science reporting in Africa. This is as a result of the increase in the number of journalists who have undertaken training on science reporting and the formation of science journalist associations in sub-Saharan Africa.
- Nevertheless, science journalists across Africa face a myriad of challenges. These include having limited skills to understand and report science, the difficult technical language used by scientists, the low interest among editors in science stories, paywalls limiting access to scientific journals, competition with other newsbeats and the perceived unpopularity of science stories.
- To enhance the coverage of scientific reports in Africa there is a need for institutions of learning to invest in the training of science journalists by integrating science journalism into their journalism curricula.
- There is a need to engage scientists to help them develop their capacities to present their scientific findings in a manner that can be easily published in mainstream media outlets.
- There is a need to encourage scientific research that can help enhance public policies, hence leading to improved standards of living in society.

Discussion questions

1. What factors influence the coverage of scientific reports in African media?
2. Suggest ways of enhancing science reporting in African media.

Suggested answers to discussion questions

Question 1
What factors influence the coverage of scientific reports in African media?

Suggested answers – summary of key points
- The perceived unpopularity of science stories – most editors and reporters think science is uninteresting and is not popular with audiences.
- Competition for space in the media – media reporting is dominated by political, sports and business news, which editors consider to be more exciting than science.
- The skills and competence of journalists – few journalists have the skills and competence to report accurate, easily understandable and interesting science stories.

- Scientific language is difficult for most journalists to understand.
- Access to sources of information – most journalists do not have access to credible sources of information, such as scientific journals.
- Specialization – journalists report better and more science stories if they are given the opportunity to specialize. Where journalists are not given an opportunity to specialize, it affects both the number and quality of science stories. Some journalists are not given the time to focus on science. As a result, they have to find extra time to report on science, which leads to fatigue.
- Emerging crises, such as climate change, epidemics and non-communicable diseases – these provide a window of opportunity because they make editors and the public more interested in science issues.
- Science journalist associations – these have contributed to the growth of science journalism in Africa.
- Development partners – donors have supported training and mentoring programmes for science journalists in Africa.

Question 2

Suggest ways of enhancing science reporting in African media.

Suggested answers – summary of key points

- Mobilizing more journalists to take an interest in science journalism.
- Introducing science journalism courses in higher institutions of learning.
- Hands-on journalism training – giving journalism skills to scientists and increasing the science literacy of journalists.
- Hiring reporters with a science background.
- Strengthening professional associations.
- Supporting journalists to access scientific journals.
- Continually reporting science stories around social problems such as epidemics and climate change.
- Creating science desks or science teams, to be the engine of science stories.
- Introducing science-related programmes in electronic media and science sections in print and online media.
- Having information officers in scientific institutions to facilitate the link between science and the media.
- Creating more interactions, such as networking events, between journalists and scientists.
- Training scientists to be able to simplify their research findings for journalists.
- Carrying out more scientific research that has direct policy and personal applications within society.
- Creating more knowledge exchange engagements between scientists and journalists: for example, conferences and the use of online networking platforms.

References

Arianrhod, R. (2019) A brief history of science writing shows the rise of the female voice. *The Conversation* 21 March. Available at: https://theconversation.com/a-brief-history-of-science-writing-shows-the-rise-of-the-female-voice-112701 (accessed 30 December 2021).

Artal, L.R. (2015) A brief history of science communication. Available at: https://blogs.egu.eu/geolog/2015/02/06/a-brief-history-of-science-communication/ (accessed 30 December 2021).

Berenbaum, M.R. (2017) Communicating about science communication: a brief entomological history. *Annals of the Entomological Society of America* 110(5), 435–438.

Jothi, P.S. and Neelamalar, M. (2010) The study of social media communication: analysis of science communication through social networking sites with special reference to scientists. Available at: https://studylib.net/doc/8271201/the-study-of-social-media-communication (accessed 30 December 2021).

Lugalambi, G.W., Nyabuga, G.M. and Wamala, R. (2011) *Media Coverage of Science and Technology in Africa*. Paris, France: UNESCO. Available at: www.unesco.org/new/fileadmin/MULTIMEDIA/HQ/CI/CI/pdf/official_documents/science_technology_reporting_africa.pdf (accessed 13 April 2019).

Massarani, L., Entradas, M., Neves, L.F. and Bauer, M.W. (2021) *Global Science Journalism Report: Working conditions and practices, professional ethos and future expectations*. SciDev.Net/CABI: Wallingford, UK. Available at: www.scidev.net/global/wp-content/uploads/Global-Science-Journalism-Report-2021.pdf (accessed 30 December 2021).

Myers, R.A. and Worm, B. (2003) Rapid world-wide depletion of predatory fish communities, *Nature* 423, 280.

Nakkazi, E. (2012) The rise of African science journalism. *SciDev.Net* 19 October. Available at: www.scidev.net/global/features/the-rise-of-african-science-journalism-1/ (accessed 12 November 2021).

National Academy of Sciences. (2017) *Communicating Science Effectively: A Research Agenda*. The National Academies Press: Washington.

Odlyzko, A. (2000) The future of scientific communication. Available at: www.dtc.umn.edu/~odlyzko/doc/future.scientific.comm.pdf (accessed 30 December 2021).

Raza, G. (2013) A brief history of science communication and 25 years of PUS in India. Available at: www.researchgate.net/profile/Gauhar-Raza/publication/235743050_A_brief_History_of_science_communication_and_25_years_of_PUS_in_India/links/09e41513084fee0dd4000000/A-brief-History-of-science-communication-and-25-years-of-PUS-in-India.pdf (accessed 30 December 2021).

Waithera, H.W. (2018) *Science in Media: Africa Science Desk Baseline Assessment Report*. Available at: www.aasciences.africa/aesa/programmes/science-communicationafrica-science-desk-asd (accessed 02 April 2021).

Zhu, Y. and Purdam, K. (2017) Social media, science communication and academic super users in the UK. Available at: www.research.manchester.ac.uk/portal/files/62968933/ASUPU.pdf (accessed 30 December 2021).

Science Journalism and Communication in Uganda: Revising a University Curriculum to Meet the Country's Needs

Dr Aisha Sembatya Nakiwala[1] and Dr William Tayeebwa[2]
[1]*Senior Lecturer, Department of Journalism and Communication, Makerere University;* [2]*Senior Lecturer, Department of Journalism and Communication, Makerere University*

In this chapter
- Context: scientists, science journalists and science communicators
- The Script programme at Makerere University
- Bridging the gap between trainers and industry
- Building a pool of reporters and communicators
- Stronger linkages between industry and academia
- Mainstreaming science journalism and communication in curricula
- Looking ahead: a brighter future for science journalism and communication in Uganda

Context: scientists, science journalists and science communicators

This chapter presents the experience of Makerere University's Department of Journalism and Communication (DJC) in mainstreaming science journalism and communication in its curricula, as viewed within the context of the current status of science journalism and communication in Uganda. It highlights insights from science journalism and communication professionals who participated in various science communication-related events at Makerere University. The experiences of students who participated in the online courses under Script are also presented.

In this chapter the term "scientist" refers to all persons who produce knowledge and innovations through scientific research. However, the concept is also used in its narrower sense to mean specialists working in various capacities in the natural, applied and physical sciences (e.g. chemistry, physics, engineering and technology). A "science journalist" is a person working for the mass media (newspapers, radio, television and online) to gather, package and distribute information produced by "scientists". The term "science communicator", on the other hand, includes all communication specialists working with scientists to disseminate scientific information.

In Africa, as in other parts of the world, scientists tend to mistrust journalists (Lugalambi *et al.*, 2011). Among their main fears is that journalists might misreport, oversimplify or sensationalize their information. Journalists, on the other hand, accuse scientists of being too technical and boring. Plausible solutions to this problem might be to either equip scientists with journalism skills so they can write stories themselves, or to train journalists to better appreciate science so they can report it better.

The challenge that exists in regard to the dissemination of scientific information through journalism and other forms of communication has been debated for years (Nakazzi, 2013). However, of particular interest for this chapter are the insights set out in a report produced by

©2022 CAB International. Science Communication Skills for Journalists: A Resource Book for Universities in Africa (Ed. Charles Wendo)
DOI:10.1079/9781789249675.0004

SciDev.Net, in which several measures were proposed to address the working conditions, practices and ethos of science journalists across the globe (Bauer *et al.*, 2013). The authors identified two worrying trends for science journalism and communication. In the first place, traditional journalism, as practised through newspapers, was in decline due to the emergence of new technologies (p. 4). With the proliferation of social media, unverified online platforms and fake news sites, the mainstream media have ceased to be the main window through which the public receive information. This poses challenges for the quality of information that the public receive. Secondly, the media are under increasing pressure arising from public relations strategies and techniques that aim to draw public attention to particular scientists, institutions and research groups (p. 5). These public relations efforts are calculated to serve the interests of particular scientists, institutions and research groups, and unsuspecting journalists can sometimes end up misleading the public due to being pressured into serving the interests of such groups. For example, in extreme cases such institutions have released exaggerated or even erroneous information about their research findings. Studies have shown that errors and exaggerations in health and science news often originate in press releases (Sumner *et al.*, 2014, 2016) or even scientific publications (Gerrits *et al.*, 2021). It takes a competent science journalist to see through these exaggerations and errors in press releases from scientific institutions.

Therefore, one of the key proposals by the authors of the SciDev.Net report was to revamp the training of journalists and communication professionals in sub-Saharan Africa and other parts of the developing world (Bauer *et al.*, 2013). This would increase the number of journalists with the competence to not only report science accurately but also to detect fraudulent scientific claims.

The Script programme at Makerere University

The 2013 SciDev.Net report laid the foundation for SciDev.Net's capacity-building programme called Script. Script is a training and networking programme for journalists, scientists and anyone who wants to communicate science in an engaging and accurate way. By connecting reporters and researchers, and giving both groups the skills to understand and communicate with each other, Script aims to increase the quantity and quality of science-related stories in the news.

Script was launched in 2018 with two free online courses, "Media Skills for Scientists" and "Science Communication Skills for Journalists" (Wendo, 2018a, 2018b).

At the launch of the Script training programme in Kigali in 2018, SciDev.Net stated that it aimed to increase "the number of articles written about science, in order to increase the application of science in public life and in the development of government policy" (Deighton, 2018).

Subsequently, more initiatives were launched by SciDev.Net, including support to various academic institutions towards the development of new curricula in science journalism and communication.

Charles Wendo, SciDev.Net's training coordinator and a past science editor at Uganda's *New Vision* newspaper, noted that the training programme would enable students to learn to report science effectively (CABI, 2018).

As of December 2021, the Script programme has supported four universities in Africa to embed science journalism and communication into their undergraduate and postgraduate curricula. The support included training-of-trainer (ToT) sessions for academic staff, technical support for curriculum development, as well as the provision of reading and teaching materials. The

Script programme also facilitated networking events that brought together journalists, communication specialists and scientists to discuss the nature of the science journalism and communication training that Uganda needed.

The Script programme activities at Makerere had three main objectives:

* ensuring that science journalism and communication are mainstreamed in undergraduate and graduate curricula
* bringing together students from journalism and communication to interact with those from the natural, biological and physical sciences
* linking faculty and students to industry professionals in science journalism and communication.

The implementation activities included induction and retooling workshops for lecturers; students' participation in Script's online courses; as well as events such as symposia and conferences where scholars and professionals shared experiences on communicating science. The programme also put in place a mentoring programme in which graduating students were paired with practising science journalists and communicators.

Bridging the gap between trainers and industry

University lecturers in the field of journalism and communication do not always have the requisite skills and competencies to address the key issues relating to scientific inquiry. For instance, what is the scientific method in knowledge creation? What are the key scientific issues of our times and how should they be packaged and disseminated by journalists? These are some of the questions that professionals pondered during two two-day workshops to induct and retool faculty at Makerere University in the art of communicating science. The first workshop, held on 4–5 October 2018, was a ToT session for the journalism and communication faculty; it was attended by eight participants. The ToT was facilitated by Dr Charles Wendo, SciDev.Net's training coordinator. He used various conceptual, theoretical and hands-on exercises to help participants appreciate the basics of science journalism and communication. The workshop aimed to plan for how best to implement the Script programme activities. A subsequent workshop was held in November 2018 to scrutinize the existing curricula and agree on how to mainstream content from the Script online courses. On 28 February and 1 March 2019, a bigger workshop was held that brought together 11 university lecturers: eight from the natural/physical/biological sciences and three from the DJC. Also in attendance were five communication professionals and five science journalists. In addition to retooling the professionals in the best practices of teaching and reporting science, the workshop also enabled networking between the university lecturers and the professionals from industry. In total, 16 lecturers attended at least one ToT session: ten were from the DJC while six came from science-based departments.

In a session on how to make science interesting to journalists and their audiences, Gerald Tenywa, an award-winning science journalist with Vision Group, pointed out the challenges journalists face in covering complex scientific issues. In the first instance, he noted the limited understanding among journalists and their editors of the concepts and principles of scientific disciplines. This is worsened by the poor facilitation available to help reporters investigate stories, especially those in hard-to-reach places. Further, a number of science stories relate to controversial topics, such as pollution, genetically modified organisms and counterfeit products, which often involve powerful organizations that are capable of influencing reporters or editors to "kill a story" or change the narrative. In addition, many science stories are dropped by editors, who find them too complex or boring, and therefore believe they will not be able to sell copy.

From a communication specialists' perspective, Anita Tibasaaga, a communications officer with the Food and Agriculture Organization of the United Nations (FAO) in Uganda, reasoned that much noteworthy science communication is about risk. She argued that science communicators must therefore be mindful of how people perceive risks and offer solutions. She also cautioned against the "deficit model" of science communication, which assumes that gaps between scientists and the public are a result of a lack of information or knowledge, ignoring other mediating factors, such as values and cultures. She called for continued engagement between science communicators, scientists, policymakers and the public who use the science.

Gerald Businge, a journalist and new media expert with Ultimate Multimedia Consult, emphasized the importance of harnessing the power of new media tools in communicating science. He noted that online platforms can help scientists to disseminate research, engage the population and keep people interested in pertinent scientific issues. Through social media, scientists can receive valuable feedback about their research findings and answer questions from the community.

Dr Wendo noted that scientific research produces vital knowledge for development, and that the media can enable that knowledge to reach the public and policymakers so they can use it to make informed decisions. However, he pointed out the shortage of journalists in Africa who can report comprehensible, interesting and accurate stories on science – a gap that the Script programme aims to address through its various interventions.

The key highlight of the February workshop was the "speed-dating" session where science journalists were paired with scientists. Every scientist had five minutes to pitch their research findings to each journalist. In the end all five journalists said they had heard one or more newsworthy research study during the session. Similarly, all six scientists said they had met and exchanged contacts with journalists they could contact in future to report a science story in the media.

An important lesson from these events is the need for continued engagement between the media industry and training institutions. The involvement of media and communication practitioners in the workshop helped university lecturers who attended to better understand the real-life challenges that the students they train are likely to face after they graduate, including:

• a lack of skills among journalists to report comprehensible, interesting and accurate stories on science
• a lack of skills among scientists and communication specialists to package scientific information that will be comprehensible and appealing to the media and the public
• a limited understanding among journalists and their editors of the concepts and principles of scientific disciplines
• poor facilitation for reporters to investigate stories, especially those in hard-to-reach places
• pressure from powerful organizations that want to influence media content
• the complex nature of science
• a perception that scientific information "doesn't sell".

This understanding of the real-life challenges that the industry faces is important when engaging in curriculum development and training, and these discussions fed into curriculum development for science journalism and communication at Makerere University.

Building a pool of reporters and communicators

A key objective of the Script programme at Makerere University was to build a robust network comprising journalism and communication students and their counterparts in the science disciplines. A good avenue for this kind of networking is the One Health network, which brings together students and faculty from across disciplines to discuss and propose solutions to health challenges. The One Health network at Makerere is part of the bigger One Health Central and Eastern Africa (OHCEA), which brings together 14 higher education institutions in six countries in the eastern and central Africa regions. The Makerere coordinator, Dr Peninah Nsamba, welcomed the Script programme's activities and encouraged students in the One Health network to participate in its online courses. As a result, 62 One Health network fellows completed the Script online course titled "Media Skills for Scientists". Meanwhile, 33 journalism and communication students completed the online course on "Science Communication Skills for Journalists" in the first cohort; eight were master's students and 25 were undergraduates. The graduating students were awarded certificates during the annual media convention held at Makerere on 25 April 2019.

Commenting on the Script online course he completed, Owen Tusiime, an electrical engineering student, said, "the course was good because it was able to teach us how to communicate our science to the general public". Irene Namyalo, a journalism master's degree student, said the skills she obtained from the course she completed were very relevant for any journalist who reports about science. Leah Oundo, an undergraduate journalism student, said the training had helped her learn how to report science: "I never thought about it before, but now I really want to be a science reporter."

The representative views expressed by the students above show that they appreciated the Script courses and considered them beneficial to their careers. It would be helpful to conduct regular evaluations with students to assess how they continue to apply the science communication skills acquired through the courses.

Stronger linkages between industry and academia

As a result of the ToT and continued engagement with Script, lecturers in the DJC developed the skills and confidence to teach, advocate for and practise science communication.

A key highlight of the Script programme at Makerere University was the April 2019 Annual Media Convention on the theme *"Communicating Science in the Social Media Age: Sharing Technical Information from Researchers with the Public"* (Makerere University, 2019). This theme was chosen to highlight the important role the media play in informing, popularizing and moderating debate about scientific knowledge. The convention attracted journalists, communication professionals, scholars and students in the field of journalism and communication, as well as participants from science/research institutions and civil society, and policymakers and government technocrats.

The convention enabled knowledge exchange between journalists and communication professionals and provided opportunities for the participants to share their experiences so as to improve and advance journalism and communication, with a critical focus on the role traditional and social media platforms play in disseminating scientific information in society.

Hon. Dr Elioda Tumwesigye, then Uganda's Minister for Science, Technology and Innovation, emphasized the importance of government and policymakers working with research and media institutions to increase the uptake of scientific information by the public (Nabatte, 2019). He emphasized the need for strong partnerships between government, donors, researchers, science institutions and the media to ensure a proper understanding of science. He stated:

The media is key to creating this link between science and society as science not communicated is science lost. Publishing findings in journals is good but the information therein never reaches the public. By reporting about scientific developments, the media contribute to public understanding of and engagement with science and technology.

In his keynote presentation, Prof. Emmanuel Dandaura, the Executive Director of the Institute of Strategic and Development Communication at Nasarawa State University in Nigeria, noted the importance of training so as to prepare the younger generation to face the challenges of tomorrow. He observed:

> We must find a way of ensuring that all our scientists in the universities are trained in science communication. We must also ensure that all our communication experts, not just the journalists or journalism students – all those that eventually end up as communication experts, including those in performing arts – we must ensure that we train them in science communication so they can use their different media to also report science and ensure that the society benefits from science.

Development communication scholar Dr Abraham Kiprop Mulwo from Moi University in Kenya highlighted the significance of science communication, which helps policymakers to make the right decisions to address the challenges that society is facing. He stated: "It is important that we begin to develop this [science journalism and communication] into a discipline. It is important that African universities begin not just developing courses but actually degree programmes that deal with science journalism."

The convention consisted of two panels, with participants drawn from academia and industry. The first panel debated the nexus between science, the media and communication for development, while the second deliberated on how to communicate science in the social media age and techniques journalists can use to disseminate technical information. Mr Daniel Kalinaki, the then convergence editor at Monitor Publications, emphasized the importance of ensuring quality reporting of science in the media. He noted:

> The world is looking for solutions and science provides a bedrock of knowledge for solving some of the problems that the world is grappling with – things like how to feed people, how to deal with ageing populations, climate change and how to fit more people in urban spaces that aren't growing.

As a follow-up to the successful Makerere University conference in April 2019, Nasarawa University organized its "2nd International Conference on Science Communication and Development in Africa" on the theme: "Humanizing Science: Optimizing Innovation and Communication for Development in Africa", from 17 to 19 October 2019 in Abuja, Nigeria. Makerere University attended the conference, with representatives from the university sharing the lessons learnt from the Script programme.

Mainstreaming science journalism and communication in curricula

The Script programme was launched at an opportune time when the DJC at Makerere University was reviewing its academic programmes. Given the strict regulations on curriculum development and implementation, the Script curriculum could not be fully adopted within the first academic year. Instead, components of the Script curriculum were adapted within 13 undergraduate and five

---

Box 4.1. Research studies carried out by postgraduate students, inspired by the Script programme.

- Interrogating the Emaciated Image of an HIV/AIDS Victim in Health Communication Messages
- The Effect of Television Programming in Preventing HIV/ AIDS among University Students: A Case study of Makerere University
- The Use of Mobile Phone Technology to Offer Bundled Agricultural Extension Services to Farmers: A Case of M-Omulimisa
- Participatory Communication and Contraceptive Use among Youth at Naguru Teenage Center
- Social Media Influence on Vaccine Uptake

---

graduate course units. As a result, 320 undergraduate students and 60 graduate students received science communication training.

It ought to be pointed out that within the newly developed master's degree curriculum, three new stand-alone courses were developed to focus on science journalism and communication: MSC 8304 – Science Communication; MSC 7202 – Health and Environment Communication; and MSC 7204 – Information, Communication and Knowledge Management. All were created under the Master of Strategic Communication programme. The programme has been approved by the University senate. The envisaged launch academic year is 2022/2023.

Another outcome of the Script programme and the online courses is that several graduate students, particularly those who participated in the Script online courses, developed research projects related to science communication. These are summarized in Box 4.1.

With specific regard to gender, 103 female students and 97 males participated in all the activities of the Script programme, particularly the online courses. Among the faculty, 11 females and nine males were retooled to teach science journalism and communication. The same faculty members were also paired with the five communication professionals (n = four females, one male) and five science journalists (n = three females and two males) (Fig. 4.1).

Evaluation of the Script programme at Makerere University

The Script programme at Makerere University involved five key activities, some of which overlapped:

1. ToT workshop for lecturers.
2. Selection of training topics for inclusion in the curriculum.
3. Approval of the curriculum.
4. Teaching of the Script curriculum.
5. Networking events.

To evaluate the success of the programme, together with the Script programme team at SciDev. Net, we set up a questionnaire online and shared it with the lecturers as well as the journalism and communication students. We received 43 responses from students. Additionally, questionnaires were delivered to lecturers who attended the ToT sessions. The results of the evaluation are summarized in the paragraphs below.

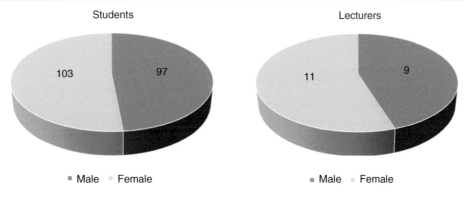

Fig. 4.1. Gender breakdown of participation in Script activities at Makerere University in 2018–2019.

Lecturers' preparedness

The first ToT took place on 4–5 October 2018, attended by eight lecturers of journalism and communication. All (100%) reported that they felt more confident to teach science communication following the training. Most of them thought the interactive discussion on teaching and packaging science communication was enriching. They felt it would be useful to have another workshop in a year to assess their progress.

The second ToT workshop took place from 28 February to 1 March 2019. It was attended by six lecturers in science courses, three lecturers in journalism and communication, eight communication specialists and five journalists. The training included a speed-dating session involving scientists and journalists. Nearly all respondents – 88% – reported that the training had given them more confidence in regard to communicating science in their different fields of work. All respondents (100%) said they had a better understanding of the general principles of science and science journalism after the training. Again, nearly all respondents – 94% – said they had learnt how to simplify science and make it more interesting to non-specialists. All participants confirmed that they had met and exchanged contacts with people they could approach for a science story in future. About 88% of respondents said that after the training they now viewed themselves as leaders in science communications. Most of them thought the speed-dating session and the session on simplifying science were the most useful parts of the workshop. They suggested allocating more time for similar workshops in future.

Students' perceptions

Overall, 93% of the respondents thought it was important for future journalists and communicators to be able to communicate science. Nearly all respondents (98%) said the course was useful to their future careers. A total of 82% were of the opinion that they had learnt to communicate science in a way that makes it interesting, relevant and accessible.

Asked what could be done to improve the content, the respondents suggested a number of ideas, including making the course more interactive (12%), introducing more practical assignments (9%) and adding more science content. Relatedly, students suggested that science lecturers be invited to teach parts of the course. Regarding delivery methods, 12% thought the training could be improved by providing students with reading materials beforehand: for example, downloadable PDFs.

Looking ahead: a brighter future for science journalism and communication in Uganda

Despite the setbacks due to COVID-19 and the closure of Makerere University for a year, the Script programme was a great success. In total, 380 journalism and communication students (n = 320 undergraduates and 60 graduates) took courses in which the Script curriculum was mainstreamed. In addition, a total of 62 students in science disciplines participated in the Script online courses and obtained certificates of completion. Overall, the Script content reached some 442 students directly, which was more than the targeted 400.

Further, the retooling workshops provided faculty with the aptitudes needed to remain focused on ensuring their students prioritize the communication of science during their professional careers. The workshops also opened channels of communication and engagement between the DJC lecturers and their colleagues in science disciplines. A mentorship scheme was also created, in which university faculty work closely with industry colleagues on pertinent issues for communication. Also noteworthy is that the Script programme has birthed within the DJC two projects funded by the Government of Uganda under the Research and Innovation Fund (RIF). The first one, headed by Dr Brian Semujju, is titled "Using Video Jockeys (VJs) to Promote Public Engagement and Awareness of Makerere University Science". The second, headed by Dr William Tayeebwa, is titled "The Multimedia Production Hub: Harnessing Digital Media for Increased Uptake and Visibility of Makerere University Research and Innovations" (Makerere University, 2020a). The two projects tie into the university's strategic plan 2020–2030, the first goal of which is to develop the institution into a "research-led university responding to national, regional and global development challenges", and the second of which is developing Makerere into "an engaged university with enhanced partnerships with industry, the community and international institutions" (Makerere University, 2020b, p. 9). It is evident that the two stated university goals echo the vision of SciDev.Net, which also emphasizes the aspect of helping individuals and organizations to apply insights from science in decision-making.

Outside of the Script programme, there are a number of science journalism associations in Uganda, such as the Uganda Science Journalists Association (USJA) and the Health Journalists Network in Uganda (HEJNU). These associations carry out training and mentorship for journalists who are interested in reporting science. In addition, media organizations are interested in science stories that are of social relevance, especially relating to health, agriculture, the environment and energy. Mirroring the media, the Uganda Journalism Awards has five categories related to science: agriculture reporting, data journalism, energy and extractives reporting, health reporting and environment reporting (ACME, 2021).

With more journalism students coming out of university with the skills to report science, more scientists getting interested in communicating their work, and interest from the media in publishing socially relevant science stories, it can be argued that the future of science journalism in Uganda is bright.

Summary

- Training of journalists, communication specialists and scientists to communicate science is necessary to increase the amount and quality of scientific information that policy-makers and the public receive through the media.
- Against this background, SciDev.Net launched the Script programme aimed at building the capacity of journalists, scientists and communication specialists to communicate science. Makerere University was selected as one of the African institutions of higher learning to participate in the programme.
- After receiving a ToT facilitated by the Script programme, the DJC at Makerere University embarked on revising the undergraduate and graduate curricula to include science communication content.
- A total of 16 lecturers – ten from the DJC and six from science disciplines – received training to enable them to teach science journalism communication.
- While curriculum review was in progress, most aspects of the Script science communication content were adapted within the existing curricula.
- A total of 380 journalism and communication students – 320 undergraduate and 60 graduate – received science communication training within the first year of the Script programme at Makerere. Another 62 students of science disciplines participated in the online courses and obtained certificates of completion.
- A key highlight of the Script programme at Makerere University was the April 2019 Annual Media Convention, which focused on science communication. Journalists, communication professionals, scholars and students in the field of journalism and communication, as well as participants from science/research institutions and civil society, and policymakers and government technocrats attended.
- The Script programme also put in place a mentoring programme in which graduating students were paired with practising science journalists and communicators.
- As a result of the skills and confidence gained from the Script programme, the DJC won two science communication projects funded by the Government of Uganda to communicate scientific research carried out at the university.
- As a result of the Script programme, several graduate students, particularly those who participated in the Script online courses, developed research projects related to science communication.
- Apart from the Script programme, there are a number of science journalism associations in Uganda that carry out training and mentorship for journalists who are interested in reporting science.
- With more journalism students coming out of university with the skills to report science, more scientists getting interested in communicating their work, and interest from the media in publishing socially relevant science stories, the future of science journalism in Uganda looks bright.

Discussion questions

1. What are some of the challenges that African journalists face in reporting on science and how would you overcome these challenges?

2. How can the introduction of science journalism training into a university curriculum contribute to national development?

Suggested answers to discussion questions

Question 1

What are some of the challenges that African journalists face in reporting on science and how would you overcome these challenges?

Suggested answers – summary of key points

- Limited understanding of scientific concepts.
- Poor facilitation for reporters to investigate stories.
- Powerful organizations that attempt to "kill" sensitive stories or influence the narrative.
- Many stories are dropped by editors, who find them too complex or uninteresting.
- Lack of skills to report accurate, understandable and interesting science stories.
- Declining sales of newspapers.
- Difficult language used by scientists.
- Mistrust by scientists, who accuse journalists of distorting information.
- Perception by some editors that science news doesn't sell.

Question 2

How can the introduction of science journalism training into a university curriculum contribute to national development?

Suggested answers – summary of key points

- Training produces journalists with the skills to report accurate, understandable and interesting science stories.
- Having more skilled journalists means more and better-quality science stories in the media.
- Information provided in science stories can help policymakers and individuals to make policy and personal decisions that can solve some of the problems that the world is grappling with.

References

ACME (2021) Uganda National Journalism Awards 2021. Available at: https://acme-ug.org/uganda-national-journalism-awards-2021/ (accessed 30 December 2021).

Bauer, M.W., Howard, S., Ramos, Y.J.R., Massarani, L. and Amorim, L. (2013) *Global Science Journalism Report: Working Conditions & Practices, Professional Ethos and Future Expectations.* SciDev.Net, London. Available at https://perma.cc/5NQM-KT6Q (accessible 30 January 2021).

CABI (2018) Makerere University joins Script training course aimed at nurturing future science journalists. Available at: https://perma.cc/9TUT-P8YU (accessed 30 November 2021).

Deighton, B. (2018) SciDev.Net launches Script training course, SciDev.Net, 28 March. Available at https://perma.cc/3YMG-QS6T (accessed 30 November 2021).

Gerrits *et al.* (2021) Reporting health services research to a broader public: An exploration of inconsistencies and reporting inadequacies in societal publications. *PLoS ONE* 16(4), e0248753. Available at: https://perma.cc/24ZP-8WLR (accessed 30 November 2021).

Lugalambi, G.W., Nyabuga, G.M. and Wamala, R. (2011) Media coverage of science and technology in Africa. Paris, France: UNESCO. Available at: www.unesco.org/new/fileadmin/MULTIMEDIA/HQ/CI/CI/pdf/official_documents/science_technology_reporting_africa.pdf (accessed 13 April 2019).

Makerere University (2019) Annual Media Convention 2018/2019. Available at: https://perma.cc/UZ7U-T9PL (accessed 30 November 2021).

Makerere University (2020a) List of Projects Awarded under RIF2. Available at: https://perma.cc/E7DZ-BZLR (accessed 30 November 2021).

Makerere University (2020b) *Unlocking the Knowledge Hub in the Heart of Africa: Strategic Plan 2020–2030*. Available at https://perma.cc/AX59-XB7N (accessed 30 January 2021).

Nabatte, P. (2019) Hon. Tumwesigye acknowledges role of media in bridging the gap between scientists and communities. Available at: https://chuss.mak.ac.ug/news/hon-tumwesigye-acknowledges-role-media-bridging-gap-between-scientists-and-communities (accessed 30 January 2021).

Nakazzi, E. (2013) What journalists want from scientists and why. *SciDev.Net* 25 September. Available at: https://perma.cc/V2MV-X5TL (accessed 30 November 2021).

Sumner P., *et al.* (2014) The association between exaggeration in health related science news and academic press releases: Retrospective observational study. *British Medical Journal* 349:g7015. 10.1136/bmj.g7015. Available at: https://perma.cc/24QH-5SM6 (accessed 30 November 2021).

Sumner P., *et al.* (2016) Exaggerations and caveats in press releases and health-related science news. *PLoS ONE* 11(12), e0168217. Available at: https://journals.plos.org/plosone/article/file?id=10.1371/journal.pone.0168217&type=printable (accessed 30 November 2021).

Wendo, C. (2018a) Media Skills for Scientists. SciDev.Net. Available at: https://perma.cc/AY9S-KGP8 (accessed 30 November 2021).

Wendo, C. (2018b) Science Communication Skills for Journalists. SciDev.Net. Available at: https://perma.cc/ZFC2-CYSP (accessed 30 November 2021).

Thinking Across Boundaries: Interdisciplinarity as the Basis of Science Journalism

Dr Samuel George Okech[1] and Dr Charles Wendo[2]

[1]*Lecturer, Department of Veterinary Pharmacy, Clinical and Comparative Medicine, School of Veterinary Medicine and Animal Resources, Makerere University, Uganda;* [2]*Training Coordinator, SciDev.Net, CABI*

In this chapter

- Introduction to interdisciplinarity
- How science is becoming increasingly interdisciplinary
- Why journalists need to understand the linkages between different scientific disciplines and sectors
- Journalism and science: shared goals, different approaches
- Why journalists need to learn some science and to think like scientists
- How journalists and scientists can collaborate for better science journalism

---

Introduction to interdisciplinarity

Ask any press-shy scientist why they are not interested in their research being reported in the media and they will express their fears that journalists will either distort or trivialize their information. Similarly, journalists who avoid reporting science stories will tell you that it takes a lot of effort to secure an interview with a scientist, and when they eventually do get the interview, the scientist uses complicated terms that the journalist does not easily understand. On the other hand, there are scientists who want their work reported in the media but too often get frustrated when they release information and find it gets no media coverage. At the same time, editors often say they find scientists' information uninteresting or too complicated for their audiences.

Whereas many people will label this state of affairs as a "conflict" or as reflecting "mistrust" between scientists and journalists, this chapter discusses it as an opportunity for interdisciplinary engagement. Journalists and scientists have different sets of skills and abilities that, when brought together, can result in better and more science stories appearing in the media. Sanza *et al.* (2019) sum it up this way: "Science journalism should be delivered by journalists who are trained in science and scientists who are trained in journalism. This will enrich the public's understanding of science from multiple perspectives and prevent blowing findings out of proportion and misleading claims from going viral."

The above quote refers to interdisciplinarity, whereby knowledge and skills from different disciplines are integrated to achieve a specific task: it could be an individual who has knowledge and skills from different disciplines, or a team of experts from different disciplines working jointly to accomplish a task. The UK Science Council, for instance, recognizes ten types of scientist (The Science Council, n.d.), of which the following seven are interdisciplinary:

- *Communicator scientists* – They combine scientific knowledge with communication skills. They work in the media, advertising and promotion, regulation and public relations.

©2022 CAB International. Science Communication Skills for Journalists: A Resource Book
for Universities in Africa (Ed. Charles Wendo)
DOI:10.1079/9781789249675.0005

- *Policy scientists* – They combine their scientific knowledge with an understanding of policy-making. They work in government departments, parliaments, advocacy groups and charity organizations.
- *Regulator scientists* – They combine scientific knowledge with competencies in monitoring and regulation. They work with regulatory bodies.
- *Teacher scientists* – They combine scientific knowledge and pedagogic skills to train the next generation of scientists.
- *Business scientists* – They combine business skills and scientific knowledge in their work in science and technology companies.
- *Developer scientists* – They combine scientific knowledge with an understanding of society to transform scientific knowledge into products or services that people can use to solve their problems.
- *Entrepreneur scientists* – They combine scientific knowledge with entrepreneurial skills and people skills to start their own business.

The science journalist needs to combine the knowledge, skills and ethical principles of journalism and science. For instance, a graduate of journalism needs to learn about the basics of scientific research, interpreting research findings, peer review, scientific consensus and the ethics of research, among other areas, to enable them to report competently on science. Similarly, a graduate of a science discipline needs to learn journalism skills and concepts, such as news values, reporting skills, interviewing techniques, structuring a story, journalism ethics and investigative journalism skills, among others. Likewise, to work effectively as a science communication specialist, an individual needs to combine the knowledge and skills of science and mass communication. In Europe, the USA and Canada, most science journalists have a degree in the sciences (Massarani *et al.*, 2021). By contrast, in Africa and Latin America, most science journalists have a journalism-related degree. It can be debated whether it is better to have a science degree and to learn journalism or to have a journalism degree and to learn about science. Whether one starts off as a journalism or a science graduate, learning the skills of the other discipline will make one a better science journalist.

Multidisciplinarity, by contrast to interdisciplinarity, involves different people using skills and ideas from different disciplines, in parallel. Actors from different disciplines address the same problem from different perspectives. This is different from interdisciplinarity, in which the actor(s) break down boundaries and integrate skills and ideas from different disciplines. This chapter focuses on interdisciplinarity.

How science is becoming increasingly interdisciplinary

While this chapter focuses on the nexus between science and journalism, it is worth noting that interdisciplinarity is also gaining ground within the sciences more broadly. Increasingly, scientists are breaking the walls between branches of science – for example, ecology and public health. Science involves multiple disciplines and its approaches have become increasingly interdisciplinary in nature. This is attributable to the increasing complexities of the challenges that require scientific interventions. Interdisciplinarity is therefore an inevitability in science today. This is clearly demonstrated by various recent happenings of global concern, such as outbreaks of diseases like COVID-19. Moradian *et al.* (2020) sum it up this way: "Never before, have so many experts in so many countries focused simultaneously on a single topic, COVID-19, and with such urgency and resolve."

A good example of interdisciplinarity is the One Health approach. One Health is defined variously but all definitions have as their central theme: collaboration between disciplines. The American Veterinary Medical Association (AVMA) defines One Health as "the collaborative effort of multiple disciplines – working locally, nationally, and globally – to attain optimal health for people, animals and our environment" (AVMA, 2008). The One Health Commission (2021) expands this definition a little, thus: "One Health is the collaborative effort of multiple health science professions, together with their related disciplines and institutions – working locally, nationally, and globally – to attain optimal health for people, domestic animals, wildlife, plants, and our environment." The One Health movement brings together specialists in human health, animal health and ecosystem health.

Stressing the need for interdisciplinarity in tackling zoonotic diseases, a *Lancet* editorial in July 2020 stated: "Safe, practical, and sustainable solutions will come from cross-specialty, inter-disciplinary, and international collaboration, not from the health and environmental sectors alone" (*The Lancet*, 2020). The editorial added:

> But operationalising a true One Health approach to prevent zoonotic diseases emerging, spreading, and harming society goes far beyond health and environmentalism. Social and political scientists, anthropologists, economists, and others must join conversations about surveillance, capacity building, and risk reduction. Industry, travel, and tourism representatives all need to be involved in changing the trajectory.

Why journalists need to understand the linkages between different scientific disciplines and sectors

Societal problems are inherently complex: changes in weather patterns; technological advancements and their associated effects; emerging and re-emerging diseases of animals and plants; changes in sociocultural norms; local, regional and global governance; economic and demographic dynamics – all of these (among others) add to the complexity of societal problems, and the associated approaches to mitigating those problems. There is no doubt that solving such problems requires interdisciplinary approaches. Scientists alone cannot provide these solutions exclusively, and indeed no single science discipline can do so, because of the varied nature and components of the interventions involved. In this chapter, we will stick with the example of health to illustrate the interlinkages between the various science disciplines and then we will highlight the invaluable complementary role of journalists in contributing solutions to societal problems.

Taking COVID-19 as a case study, it is already obvious that epidemiologists, physicians, laboratory technologists, veterinarians, virologists, biochemists, nurses, immunologists, chemists, pharmacists, dispensers, sociologists and many other categories of scientists have been working, day and night, to tackle the pandemic. Each of these disciplines makes a unique contribution to this fight that the others would either not be able to make at all, or could not do satisfactorily. The effects of one, however stellar, would not deliver the solution. The output of each is either a building block on which others add a layer or a tool for the others to use to progress towards the solution, or to make their unique contribution. Knowledge generated by one discipline either triggers a process of further knowledge creation or is applied directly by another to solve a problem.

However, while scientists in different disciplines can usually communicate easily and freely with each other, their language and content are largely unpalatable to the public, who, in many cases, are the ultimate users of the information. For example, consider the following hypothetical statement depicting what a scientist might say about the virus that causes COVID-19: "Severe acute respiratory syndrome coronavirus 2 (SARS-COV-2) is a ribonucleic acid virus with a phospholipid bilayer envelope that can be disintegrated by either 70% alcohol or a surfactant – such as a detergent – through its hydrophobic properties …".

Such a message might not be understandable to the public and yet there is information in it that they need to apply to protect themselves and prevent the spread of the disease. The information thus needs to be translated.

The above message, packaged in heavy scientific jargon, can be stated in simple terms as follows: "The coronavirus that causes COVID-19 disease dies when 70% alcohol or foam from soap is applied to it." This message is the basis for recommending the use of 70% alcohol-based hand sanitizer and soap for hand-washing to eliminate the virus. Science communication experts or journalists can easily bridge this gap in communication by repackaging such scientific statements as information that the public can understand and relate to. Indeed, reflecting on the measures taken since the early stages of the pandemic, not much would have been achieved without the contribution of journalists: they interacted with scientists, picked the information, repackaged it appropriately and disseminated it to the public in a palatable and usable form. In many instances, through professionally managed interviews, they extracted information from scientists on behalf of the public. Through their questions, they were also the voice of the voiceless – a huge section of the public that would otherwise not have been able to ask scientists to answer their questions. They ensured that information reached every section of the public through the media. They also mobilized the public and kept them up to date on standard operating procedures set by the WHO and national governments (informed by scientific evidence), through social media, blogs, radio, television and print media channels.

When doing their reporting, journalists need to take into account the fact that many problems are complex in nature and do not have one-track solutions. A solution will involve many actors and disciplines. As the complexity of the problem increases so does the need for more actors, and the efforts thereof. Maxwell (2001) summarizes this in what he calls the "law of Mt Everest": "as the challenge escalates, the need for teamwork elevates". Take, for example, the jigger (*Tunga penetrans*) epidemic that once broke out in humans in eastern Uganda: tackling it required medical doctors, veterinarians, environmental scientists, nurses and social scientists. The implication of this for journalists is that they need to interview/interrogate sources from different disciplines when reporting on an issue. Understanding the complex nature of problems also helps journalists to figure out who to target with which part of their story.

In summary, journalists must pay attention to the different disciplines and sector players involved in research and intervention projects. For example, reporting on COVID-19 as if it were a purely medical problem does not suffice. Moradian *et al.* (2020) describe the role that the medical sciences, social sciences and biological sciences, as well as information technology and mathematics, all play in tackling the COVID-19 pandemic. Journalists need to pay attention to these various dimensions and the links between them: for example, how imposing a quarantine might affect other human concerns, such as mental health and people's incomes. Conceptualizing the different dimensions of a problem will help the journalist to write a complete, multi-sourced, well-corroborated story. For instance, a story on a new high-yielding variety of a staple food would not be complete without mentioning whether or not the new variety is socially acceptable, and why.

The journalist thus has their job cut out for them, to integrate perspectives from different disciplines and sectors in order to write a balanced and accurate, highly informative story about a scientific development.

Journalism and science: shared goals, different approaches

The role of the research scientist is to conduct studies by applying the scientific method (Andersen and Hepburn, 2019) to generate knowledge that provides solutions or links to solutions to societal problems, through the use of scientific approaches. The scientist also has an obligation to disseminate the knowledge they generate through their studies. It is expected that the knowledge generated through scientific research will directly or indirectly contribute to improving society. Ultimately, research is all about people. The journalist's role, on the other hand, is to deliver timely, accurate and well-packaged information to the right audience. Both journalists and scientists aim to provide evidence-based information that will enhance the applied knowledge available in society and eventually improve the livelihoods of the community. Ultimately, journalism is also all about people. Fig. 5.1 makes clear that journalists and scientists need each other; ideally, both work together to benefit society.

However, journalists and scientists differ quite a bit in their approaches to their respective jobs. Unlike journalists, who are trained to present the main point of the story in their opening statement(s), scientists are trained to patiently tell a story through its stages of development before stating/arriving at the conclusion. Scientists follow and assess ideas, actions and results at the various stages of their development. While they do this, they are searching for evidence and they wish to clearly present the argument before they finally reach the conclusion. Scientists very rarely "front-load" a story. Unlike journalists, who are relatively more constrained by the turnaround time of a story and by space availability – which both affect the scope of the story – scientists are more accustomed to a relatively larger space for expression and a longer time to dwell on the story. They also target different audiences with their messages.

Considering the above differences, can scientists and journalists work together effectively? The answer should be an emphatic yes. They ought to complement and support each other. This starts with understanding the two different worlds they live in, with reference to their respective approaches and styles.

Both professions require a high degree of curiosity. Scientists, especially those involved in research, are curious: they wish to know/learn something new and have a great desire to create or innovate solutions to societal problems. This curiosity generates excitement about a given subject. To be engaged effectively, therefore, their curiosity must be aroused, often by encountering challenging questions that point to the hitherto "unknown", either asked by themselves or

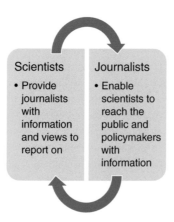

Fig 5.1. Why journalists and scientists need each other.

raised by others. Scientists also love to share their new findings, but they tend to think their audiences are "patient" listeners who can first be taken through the detailed background and processes before the conclusion is stated. The question of the selection and prioritization of information to share and how to package it pops up at this stage – what the scientist considers important may not necessarily be interesting to and/or immediately usable by the journalist and the public; hence the need for scientists and journalists to cooperate.

Why journalists need to learn some science and to think like scientists

Journalists ought to learn and/or constantly refresh their basic knowledge of science. This is important in order to strengthen their appreciation of and ability to easily follow the steps through which scientific processes pass to deliver solutions/impact. For example, a research process for the production of a COVID-19 vaccine goes through several iterative processes on computer, in the laboratory and in animals before eventually ending up in humans – from the science laboratory to the arm of the vaccine recipient. Armed with this information, a journalist will be able to report about COVID-19 vaccine research more accurately. Journalists should also familiarize themselves with key scientific terms so as to be able to decipher a scientist's message. Quite often, scientists do not know how to convert their jargon-laden messages into popular versions for public consumption. The journalist should therefore be equipped to simplify complicated information for public consumption – importantly, without changing the meaning (scientists hate to be misreported, misinterpreted or misquoted). Journalists need a basic understanding of the scientist's work in order to ask them the right questions and to make sense of their research findings and their implications. In this way, science journalism can overcome the various criticisms made of science journalists highlighted by Secko *et al.* (2013), including hyping research findings, uncritical reporting, emphasizing frames of scientific progress and economic prospects, not presenting a range of expert opinion, having a preference towards positive messages and reporting unrealistic timelines. Journalists also need to know how to read a scientific paper, make sense of it and spot the news in it. To do this they need to know how scientists structure their research papers, the different types of research and their limitations, and the key terms used in scientific research and their meanings.

How journalists and scientists can collaborate for better science journalism

The scientist's world, as indicated earlier, is one of curiosity. Scientific researchers are curious about understanding the challenges at hand and possible solutions to societal problems. The world of scientific research is one of meticulous step-by-step processes that call for patience, especially on the part of journalists and the public. This is contrary to the journalist's world, where the main point in a message is front-loaded. The scientist's world is also heavily jargon-laden. Scientists therefore have a lot to learn from journalists regarding how to reach the public with their information. They need to learn the ability to translate sophisticated scientific jargon, formulas, equations and processes into short, simple and attractive messages for lay persons. The story should not only be informative but also captivating and memorable. Furthermore, journalists must learn to live with some "uncomfortable" demands by scientists, including embargoes, procedures for getting interview appointments and no-go areas in scientific establishments. Eise

(2019) recommends embedding social scientists in hard science research teams as a method of enhancing institutional science communication. This could work for journalists too.

Journalists and scientists should collaborate to ensure better science journalism. While this can be learnt on the job, it is a good idea to cultivate such a culture of collaboration early on in the careers of the two parties. Training institutions, where the basic concepts and principles can be taught and learnt, present the most fertile ground for cultivating this culture. In this regard, Nasarawa State University Keffi in Nigeria and Makerere University in Uganda have carried out networking events to enable journalism and communication students to interact with their counterparts in science disciplines. These networking events have taken different forms, such as, for example, the 2019 Annual Media Convention at Makerere University (CABI, 2019).

For practising journalists, programmes such as the Knight Science Journalism Fellowship Programme at the Massachusetts Institute of Technology (MIT) offer an opportunity to engage with science. The fellowship helps to deepen their understanding of science and build their skills for reporting science. Furthermore, a number of organizations, such as the African Science Literacy Network in Nigeria (ASLN) and the Science Foundation for Livelihoods and Development (SCIFODE) in Uganda, have arranged scientist–journalist pairing schemes where the two are given an opportunity to work with each other for a defined period. Other approaches that have been used to facilitate journalist–scientist engagements include joint field visits, internships for journalists at scientific institutions and internships for scientists in the news media.

Not surprisingly, a 2011 study showed that relations between journalists and scientists in Africa were improving (Lugalambi *et al.*, 2011, p. 18). This study involved content analysis of science stories in six African countries and in-depth interviews with journalists. The journalists interviewed for the study reported an improvement in their relations with scientists, largely because both sides had started appreciating each other's relevance in their work: journalists need scientists as sources of information and comments for their stories; scientists realize that their information will be of limited benefit to the public if it is not disseminated through the media. Similarly, in the Global North, despite occasional disputes that occur when journalists misquote scientists or report a story that paints them in a negative light, most scientists are positive about interaction with the media (Peters, 2013). Nevertheless, most scientists reported wishing they had more control over the information published by journalists: for example, they would like the opportunity to check the final version of the article that journalists write after interviewing them (which most journalists are unwilling to accept).

It can be argued therefore that if journalists and scientists wish to jointly move in the right direction they ought to recall Lencioni's (2002) analogy of what an organization needs to do if it wishes to excel in its trade: "If you could get all the people in an organization rowing in the same direction, you could dominate any industry, in any market, against any competition, at any time."

Discussion questions

1. Discuss why it is important to interview scientists from different disciplines when reporting on the pollution of an important water body in your country.
2. Discuss why it is important for journalists and scientists to network.

Summary

- No major societal problem can be solved by a single profession. Due to the multifaceted nature of problems, scientists are increasingly breaking professional walls and collaborating with other professional disciplines.
- Journalists need to embrace interdisciplinarity in one or more of the following three ways.
  - They need to be aware of the multidimensional nature of problems. This will help them shape the scope of their stories, and map out who to interview and where to find additional information. This will result in more contextualized and more comprehensive stories.
  - They need to build good working relations with scientists. Science journalism is an inevitably interdisciplinary endeavour that cannot be accomplished successfully without working with scientists.
  - They need to learn the basics of science and scientific research in order to report accurate, comprehensible and relatable science stories.
- Interdisciplinary efforts yield much better results than an individual discipline can achieve on its own. Interdisciplinarity works wonders. Likewise, reporting with an interdisciplinary mind yields more contextual science stories.

Suggested answers to discussion questions

Question 1

Discuss why it is important to interview scientists from different disciplines when reporting on the pollution of an important water body in your country.

Suggested answers – summary of key points

- Social problems, such as pollution, have different aspects – for example, health, ecological, agricultural, nutritional and economic dimensions.
- The multidimensional nature of the problem requires expertise from different disciplines.
- Increasingly, scientists are using an interdisciplinary approach to solve social problems. No single specialty can provide all the answers. This makes it possible for a journalist to find scientists of different disciplines to interview for the same story.
- Interviewing scientists from different disciplines helps the journalist to write a more complete story. If they rely on the expertise of only one discipline, a journalist could miss out on important perspectives on the story.
- Integrating perspectives from different disciplines helps the journalist to write a more informative, balanced and accurate story.
- In summary, by incorporating knowledge and views from different areas of expertise, journalists provide the public and policymakers with a better understanding of societal issues.

Question 2

Discuss why it is important for journalists and scientists to network.

Suggested answers – summary of key points

- Journalists need scientists as sources of views and information.
- Scientists need journalists to reach the public and policymakers with evidence-based information, which they can use for making policy, administrative or personal decisions.

- Journalists and scientists have different approaches. Networking helps the two sides to understand each other better, which leads to better and more science stories in the media. Additionally, it enhances the capacity of journalists to effectively moderate media discussions on scientific or science-related topics, as much as it helps scientists to improve their capacity to communicate with the public on such fora.
- By networking with scientists, journalists have more opportunities to learn about developments in the scientific world and their implications for society.
- Collaborating with scientists gives journalists more opportunities to access scientific sites, such as laboratories. It also reduces the degrees of separation between individual journalists/media houses and individual scientists/science establishments.

References

Andersen, H. and Hepburn, B. (2019) Scientific method. In Zalta, E. (ed.) *The Stanford Encyclopedia of Philosophy*, summer 2019 edn. Available at: https://plato.stanford.edu/archives/sum2019/entries/scientific-method/> (accessed 21 March 2021).

AVMA (2008) One Health: a new professional imperative. Available at: www.avma.org/KB/Resources/Reports/Documents/onehealth_final.pdf (accessed 9 March 2021).

CABI (2019) Script project paves the way for Uganda's first Annual Media Convention dedicated to science journalism. Available at: www.cabi.org/news-article/script-project-paves-the-way-for-uganda-s-first-annual-media-convention-dedicated-to-science-journalism/ (accessed 14 December 2021).

Eise, J. (2019) What institutions can do to improve science communication. *Nature*. Available at: http://www.nature.com/articles/d41586-019-03869-7 (accessed 20 May 2021).

Lencioni, P. (2002) *The Five Dysfunctions of a Team: A Leadership Fable*, 1st edn. Jossey-Bass, San Francisco, California, USA.

Lugalambi, G.W., Erastus, E., Lambiv, G.T., Lukanda, I.N., Nyabuga, G.M., *et al.* (2011) Media Coverage of Science and Technology in Africa. Available at: http://www.unesco.org/new/fileadmin/MULTIMEDIA/HQ/CI/CI/pdf/official_documents/science_technology_reporting_africa.pdf (accessed 23 June 2021).

Massarani, L., Entradas, M., Neves, L.F. and Bauer, M.W. (2021) *Global Science Journalism Report: Working conditions and practices, professional ethos and future expectations*. SciDev.Net/CABI, Wallingford, UK. Available at: www.scidev.net/global/wp-content/uploads/Global-Science-Journalism-Report-2021.pdf (accessed 30 December 2021).

Maxwell, J.C. (2001) *The 17 Indisputable Laws of Teamwork: Embrace Them and Empower Your Team*, 1st edn. Thomas Nelson, Nashville, Tennessee, USA.

Moradian, N., Ochs, H.D., Sedikies, C., Hamblin, M.R., Camargo Jr, C., *et al.* (2020) The urgent need for integrated science to fight COVID-19 pandemic and beyond. *Journal of Translational Medicine* 18, 205. https://doi.org/10.1186/s12967-020-02364-2.

Peters, H.P. (2013) Gap between science and media revisited: Scientists as public communicators. *PNAS* 110 (Supplement 3), 14102–14109. Available at: www.pnas.org/content/pnas/110/Supplement_3/14102.full.pdf (accessed 28 June 2021).

Sanza, C., Borowiec, B.G., Secko, D., Qaiser, F., Ferreira, F.A., *et al.* (2019) Why we see hope for the future of science journalism. *The Conversation*. Available at: https://theconversation.com/why-we-see-hope-for-the-future-of-science-journalism-111244. (accessed 24 June 2021).

Secko, D.M., Amend, E. and Friday, T. (2013) Four models of science journalism. *Journalism Practice* 7(1), 62–80, DOI: 10.1080/17512786.2012.691351.

The Lancet (2020) Zoonoses: beyond the human–animal–environment interface. *The Lancet* 396(10243). Available at: www.thelancet.com/action/showPdf?pii=S0140-6736%2820%2931486-0 (accessed 30 December 2021).

The One Health Commission (2021) What is One Health? Available at: www.onehealthcommission.org/en/why_one_health/what_is_one_health/ (accessed 19 May 2021).

The Science Council (n.d.) 10 types of scientist. Available at: https://sciencecouncil.org/about-science/10-types-of-scientist/ (accessed 14 December 2021).

The Place of Science in the African Newsroom

Dr Darius Mukiza

Journalist, lecturer, Coordinator of Post-graduate Studies and Head of the Public Relations and Advertising Unit at the School of Journalism and Mass Communication, University of Dar es Salaam

In this chapter
- Introduction to science in the media
- Why the media should report science
- Media reporting of science in Tanzania
- What kind of science news do editors want?
- What information should a journalist seek from a scientist?
- Why a science journalist should not become a mouthpiece for scientists

Introduction to science in the media

Analyses of science journalism in Africa show both positive and negative trends. On the one hand, a report published by the African Academy of Sciences painted a picture of "a dying profession", largely due to the closure of science desks in the mainstream media (Waithera, 2018). This "dying profession" narrative mirrors a trend that has been observed in the Global North, where media houses are closing down their science desks and laying off science journalists as a result of business pressures, which demand that media companies make money – leading to them prioritizing the types of stories that give them better sales. For example, the number of dedicated science sections in US newspapers declined from 95 in 1989 to 19 in 2012 (Morrison, 2013).

However, shutting down science sections and science desks does not necessarily imply an end to science stories. Instead, science stories might be written by generalist journalists and edited by generalist editors, rather than science editors. The stories would then be published in sections that are not dedicated to science. However, such changes can lead to media organizations publishing fewer and lower quality science stories.

On the other hand, global science journalism reports (Bauer *et al.*, 2013 and Massarani *et al.*, 2021) present an optimistic view of the future of science journalism in the Global South, despite various challenges. They show that most science journalists disagree with the narrative that science journalism is a dying profession. The reports point out that the Global South has younger science journalists than the Global North, probably indicating a higher number of newcomers to the profession. Relatedly, a feature article published by SciDev.Net indicates that there was an increase in the number of science journalists in Africa as early as 2012 (Nakkazi, 2012), which was attributed to donor-supported training and mentorship opportunities, as well as the activities of science journalism associations.

©2022 CAB International. Science Communication Skills for Journalists: A Resource Book
for Universities in Africa (Ed. Charles Wendo)
DOI:10.1079/9781789249675.0006

Why the media should report science

The role that journalists can play in disseminating scientific information and enhancing public understanding of science and technology is well summarized in the UNESCO report *Media Coverage of Science and Technology in Africa* (Lugalambi *et al.*, 2011, p. 5): the media provide the public with information they use to comprehend the world they live in and to make informed decisions. Thus, media coverage influences the way people understand science and technology and the world around them.

The world is built on science; it is hard to think of an area of modern social and professional life that is not impacted by science. For example, the influence of social media, the innovation of modern trains and planes, and the cars we use to drive to work are products of scientific research. Furthermore, many of the development challenges facing African countries such as hunger, disease and land degradation have their solutions in science. Thus, the media should report science because science influences people's daily lives.

According to a technical report issued by the Expert Working Group on Science and the Media in Australia, the media should report science because it has a huge role to play in informing how we respond to many of the challenges we face as a society (Inspiring Australia, 2011). Indeed, with the coming of COVID-19, the entire world is looking to science to provide effective treatments, vaccines and knowledge about how to keep safe.

As Clifford *et al.* (2009) pointed out, in a democratic society the media serve one or more of the following roles:

- Monitorial role: scanning the environment for information that is relevant to citizens, thereby meeting the audience's information needs.
- Facilitative role: providing a platform for citizens to express their views and possibly influence governance decisions on matters that affect them.
- Radical role: giving citizens an opportunity to challenge those in power and oppose injustices.
- Collaborative role: working in a symbiotic relationship with their governments to address societal problems.

Reporting science in the media can serve any of the above four roles. It leads to more awareness about science and technology and can help people make informed policy or personal decisions. It can also lead to more people taking interest in science and technology careers. The media can allow the public to "own" science and engage in debate on it, as much as they own and engage with sports, music or politics, by treating science stories in such a way as to stimulate debate and discussion. The media have great power when it comes to moulding public attitudes on a wide range of scientific issues and they should not abdicate this role.

Media reporting of science in Tanzania

In preparation for writing this chapter, in March 2021 the author interviewed ten Tanzanian media practitioners to get qualitative information on how science was reported in their media outlets. They included three editors, three senior journalists and four early career reporters, all from the print media (*Mwananchi* newspaper, *The Citizen* newspaper, *Habari Leo* newspaper, *The Hill Observer Laboratory* newspaper and *Nipashe* newspaper). They were asked questions regarding the type of science stories editors prefer, what kinds of questions they would ask a scientist and how journalists can remain independent of scientists when reporting science stories.

Their views are discussed here with reference to the available literature on the subject and the author's experience as a practising journalist.

No science desks

The media practitioners all said their media organizations did not have science desks, and there was limited reporting of science in the news. Editors were said to give more priority to business, sports and political stories.

Lugalambi *et al.* (2011) point out a number of reasons for the low coverage of science in African media. Firstly, most journalists find scientific topics too complicated. Secondly, journalists in Africa often lack the knowledge and skills to report effectively on science stories. Thirdly, media organizations make insufficient investments towards improving the capacity of their journalists to report science. Fourthly, most scientists do not trust that journalists will report their information accurately: they fear they will be misquoted or reported out of context. As a result, science stories are overshadowed by more sensational topics, such as politics, sports and business.

However, there is hope that these challenges can be overcome, especially as a number of African institutions of higher learning have now introduced science journalism and communication training into their curricula, with the support of SciDev.Net's Script programme. For example, in what is a first in Tanzanian journalistic history, the School of Journalism and Mass Communication (SJMC) at the University of Dar es Salaam had trained more than 170 journalism students and 14 lecturers on science journalism as at March 2021. Moreover, a science journalism course has been embedded in the SJMC curriculum. This is a good start towards promoting science journalism in African newsrooms.

What kind of science news do editors want?

All media outlets have to meet the basic minimum requirement of surviving as a business or not-for-profit organization, which requires that their output remains appealing to a critical mass of listeners, viewers and readers. Thus, editors require that their stories should be useful and relevant to their audiences, and timely. When selecting their line-up of stories to run, editors will often discard those they think their audiences are not likely to be interested in. In this context, the interests of editors tend to mirror those of their target audiences: the most commonly reported fields of science in the African media include the environment, health, technology, and food and nutrition (Lugalambi *et al.*, 2011, p. 23). These are the science topics that speak to people's daily needs and aspirations. However, these priorities can change due to contemporary issues: for example, the outbreak of COVID-19 led to a sudden increase in health stories globally.

In a practical guide for journalists published by SciDev.Net, a former science editor of the UK *Guardian* explained that the public appreciate science stories that relate to their own experiences (Radford, 2008):

> Most people are not interested in "science". They are interested in what they eat, what makes them sick, why they feel miserable, how they get better and what makes them die.
>
> They do not, as far as they are aware, want to know about general advances in virology, neuroscience, drug design or gerontology. But you can get them interested in the antioxidant properties of Chilean wine, the killer tactics of the HIV and Ebola virus, or why some musicians have perfect pitch. This has nothing to do with being anti-science. It has to do with people's preference for the particular, rather than the general.

Journalists should cover issues in which science impacts on society and policy. That entails looking at areas such as climate change, energy, pollution, genetically modified crops, medicine, nuclear proliferation, natural disasters and other pressing issues that arise from time to time.

From the interviews with journalists carried out in preparation for writing this chapter, we can identify four things that editors of mainstream news media want in science stories.

Human interest stories

Editors prefer human interest stories in every beat of their news coverage. This means that even science-oriented stories should have some element of human interest: they should contain both real-life experiences and scientific facts related to those experiences. Human interest storytelling is one of the most effective ways to get people interested in science stories. Embedding human interest elements also helps media audiences to understand and appreciate science better.

Relevance to people's lives

The journalists who were interviewed also said that editors prefer stories that have a direct impact on people's lives. For instance, in Tanzania most people rely directly or indirectly on agriculture as their main source of livelihood. Therefore, a scientific development that improves agricultural productivity is newsworthy. This includes stories on the mechanization of agriculture, new farming techniques that improve productivity, new and superior crop varieties, value addition to agricultural produce, and new agricultural initiatives that create more jobs – especially for the young. Another journalist said that their preferred science stories are those related to human health, livestock health and economic activities, because these issues directly affect people's livelihoods. One journalist stated: "My editors would go for those stories that lead to solving people's problems in the community." Another journalist stated: "My editors would be more than happy to publish stories that revolutionize agriculture."

Solutions to people's problems

It also came out from the interviews that editors want stories that show a clear solution to people's most pressing problems, such as health problems. One respondent explained:

> Medical science plays a pivotal role in keeping the society healthy, engaging it in the production chain and contributing to economic growth; hence a healthy people is a healthy nation. Therefore, editors would like to cover the health story particularly on how a new development provides a relief to health problems and leads to improvement of people's living standards.

As pointed out by Newman *et al.* (2019) in the *Reuters Institute Digital News Report 2019*, globally a growing number of people are avoiding the news because of the abundance of negative stories, which negatively affect their mood. The report states: "This may be because the world has become a more depressing place or because the media coverage tends to be relentlessly negative – or a mix of the two." This problem can be partly addressed by publishing more stories that offer solutions to the challenges that the world is grappling with, such as poverty, health and hunger. Not surprisingly, therefore, editors are more interested in scientific developments and topic areas that offer solutions to people's problems.

Connection to trending discussions

Editors' choices are usually dictated by the editorial needs at that moment in time: for example, they will find a science story related to a trending topic more attractive. When a major story is trending in social discussions, editors tend to publish stories related to that topic. For instance, if there is a major outbreak of a disease, editors are likely to publish a science story related to that disease. Similarly, in the case of a major disaster, editors are likely to accept a science story that helps the public to make sense of the disaster and overcome or mitigate it. Every trending story thus offers an opportunity for journalists to report a science angle on the story. Science journalists need to be alert, observe their environment and spot the window of opportunity in every trending story.

Relatedness to editors' preferences

Most respondents who were interviewed agreed that editors' preferences vary. Some editors are more inclined than others to publish science stories. If an editor has a preference for science stories, they are likely to opt for news that explains science, but that frames it in a way that catches the reader's attention. Some editors prefer science stories with political, business or entertainment angles. Others prefer stories that involve some controversy. On the other hand, if the editor is not inclined towards science, they are not likely to give space to science stories at all. Nevertheless, a science story could appeal to them if it has political, disaster or economic dimensions.

Understandable, informative, authentic and accurate

Above all, every editor wants a science story to be informative, authentic and accurate, and also to be written in simple language that their audiences will understand, without distorting the meaning.

In a nutshell, most of the time, editors' choices are dictated by the editorial needs at that time, so it depends very much on how the reporter pitches the science story as to whether it catches the editor's attention.

What information should a journalist seek from a scientist?

As in any field of journalism, editors prefer stories that answer what is commonly known as the "5Ws + H": what, where, who, when, why + how. Thus, when reporting on the findings of scientific research, a reporter should ask the following questions.

- *What* – What is the most important research finding? What is new about it? What is most surprising about it? What are the implications of that finding? What should be done about it? What next (more studies, a new community intervention, etc.)?
- *Where* – Where was the research carried out? Where (in which journal) were the results published? Where (at which scientific meeting) were the results presented?
- *Who* – Who were the scientists that carried out the research? Who do the research findings affect? Who funded the research?

Box 6.1. Example of a science story that has an impact on society.

In 2017, an editor at the *Guardian on Sunday* of Tanzania, the late Mbena Mwanatongoni, asked reporters to suggest story ideas that would impact on society and stimulate action by the authorities. One journalist successfully pitched a story on the impact of jiggers in one region in Tanzania. A matrix for covering that story was drawn up, which included interviewing the victims, the authorities, medical doctors and respondents from schools, and reviewing scientific reports. When the story was published, it was well received by the public because it provided useful scientific information on a problem affecting them. Subsequently, the authorities in the region concerned took action on the jigger problem. This is the kind of story an editor will readily publish in their newspaper.

- *When* – When was the research carried out? When were the results published?
- *Why* – Why was it necessary to carry out this research? Why should the public care about the findings? Why is it important now?
- *How* – How was the research done? Based on the findings, how should society change (for example, implementing new policies or engaging in behaviour change)?

Similarly, Scott (2012) proposes a number of standard questions that always work for journalists when interviewing scientists. These are particularly useful for event-based stories such as natural disasters.

- What is the occasion?
- What and how big is the problem?
- What is the cause?
- What are the consequences?
- What is the solution?
- What is its effect?

The journalists who were interviewed for this chapter were of the view that their role is to ask questions on behalf of the public. Questions should thus be related to what the public are going to be interested in hearing about. For example, the public want to know how safe and effective a new COVID vaccine is, whether it has side effects, whether it has been approved, and how available it is; most will not be interested in the scientific details such as the type of viral protein that the vaccine is based on. If the scientist uses terms that the public might not understand, the journalist should ask them to explain those terms. Indeed, journalists are a bridge between scientists and those we call the beneficiaries of scientific research. They should endeavour to explain complex scientific issues in simple language.

Generally, before interviewing a scientist, journalists should try to think like their readers. This will enable them to ask questions about the key things that their audiences would want to know about.

Why a science journalist should not become a mouthpiece for scientists

The *Code of Ethics* for the Society of Professional Journalists (SPJ) urges journalists to act independently by avoiding bidding for news and conflicts of interest: "The highest and primary obligation of ethical journalism is to serve the public" (SPJ, 2014).

This implies that a journalist must remain objective and not succumb to undue influence from scientists when reporting on science issues. They should serve the interests of the public rather than those of scientists. The Tanzanian journalists interviewed for this chapter agreed with the view that journalists should act independently at all times, even though they need the cooperation of scientists to report good science stories. One said: "Journalists are there to report and explain scientific issues to the targeted audience, rather than being a brainwasher."

The editors who were interviewed said that a science journalist should not become part of the scientific fraternity because that would compromise their objectivity. They insisted that journalists should try to state everything as it is, and that they must explain things in simplified language.

Moreover, the editors who were interviewed said that journalists are expected to analyse and question scientists' opinions where necessary, and not merely be used as a conveyor belt of information. One editor said that some scientists can have hidden motives or can have opinions that contradict the generally accepted scientific position about an issue. In such a case, journalists should corroborate information with other sources, to avoid being misled by one scientist. To clarify his point, the editor said that when reporting on particular research findings, journalists need to add the views of other experts in the same field. This will help to ensure objective reporting. In sum, a journalist should not simply convey one scientist's information without doing any research and asking questions.

Summary

- Science journalism in Africa is not a dying profession. The African continent has a growing crop of, often young, science journalists.
- However, most media organizations do not have a science desk. This means that most science stories are handled by generalist reporters and editors, and are published or aired in non-specialized sections or segments of the media. This has implications for the quality and quantity of science stories.
- Not surprisingly, stories on politics, business, sports, entertainment and other sensational topics dominate the media, while science stories receive little coverage.
- It is useful for science journalists to know that editors in the mainstream media want science stories that have human interest elements, are relevant to people's lives, offer solutions to people's problems, or are connected to a trending social discussion, and are easy to understand. Such stories address the needs of the public and are likely to attract audiences to the media outlet concerned while providing the public with important information.
- While scientists are important for journalists as news sources, every science journalist needs to act independently and to avoid being subject to undue influence or manipulation by scientists.
- Science journalists have the important role of providing the public with useful information that they can use when making decisions.
- While the number of science journalists is growing, a lot still needs to be done to improve the number and quality of science stories in the media.

Discussion questions

1. Explain in detail why science journalism should be prioritized in African countries.
2. How can science journalism be made more sustainable in African countries?

Suggested answers to discussion questions

Question 1
Explain in detail why science journalism should be prioritized in African countries.

Suggested answers – summary of key points
- Science is the foundation of development.
- Through the media, science knowledge can reach politicians and the public so that they can use it to make better policies and personal decisions.

- Science affects people's lives.
- For most people in society, the media is their main source of science information. What the media do not report about, they might never know.

Question 2

How can science journalism be made more sustainable in African countries?

Suggested answers – summary of key points

- By establishing science journalism desks.
- By providing training in science journalism (both in-house and external training).
- By collaborating with scientists.
- By preparing debates on science journalism.
- By networking with science journalism stakeholders.
- By developing and making science journalism a core course.
- By training lecturers in science journalism.
- By researching and publishing science journalism.
- By training media stakeholders in science journalism.
- By establishing science journalism clubs among students.

References

Bauer, M.W., Howard, S., Ramos, Y.J.R., Massarani, L. and Amorim, L. (2013) *Global Science Journalism Report: Working conditions and practices, professional ethos and future expectations.* SciDev.Net, London. Available at https://perma.cc/TF2W-3YH5 (accessed 30 January 2021).

Clifford G. C., Theodore L. G., Denis M., Kaarle N. and Robert A. W. (2009) *Normative Theories of the Media: Journalism in Democratic Societies.* University of Illinois Press, Urbana, Illinois.

Inspiring Australia (2011) *Science and the Media: From Ideas to Action.* Available at: www.industry.gov.au/sites/default/files/2018-10/inspiring_australia-science_and_the_media-from_ideas_to_action_2011.pdf (accessed 24 December 2021).

Lugalambi, G.W., Nyabuga, G.M. and Wamala, R. (2011) *Media Coverage of Science and Technology in Africa.* Paris, France: UNESCO. Available at: www.unesco.org/new/fileadmin/MULTIMEDIA/HQ/CI/CI/pdf/official_documents/science_technology_reporting_africa.pdf (accessed 13 April 2019).

Massarani, L., Entradas, M., Neves, L.F. and Bauer, M.W. (2021) *Global Science Journalism Report: Working conditions and practices, professional ethos and future expectations.* SciDev.Net/CABI, Wallingford, UK. Available at: www.scidev.net/global/wp-content/uploads/Global-Science-Journalism-Report-2021.pdf (accessed 30 December 2021).

Morrison, S. (2013) Hard numbers. *Columbia Journalism Review.* Available at: https://archives.cjr.org/currents/hard_numbers_jf2013.php (accessed 23 June 2021).

Nakkazi, E. (2012) The rise of African science journalism. SciDev.Net, 19 October. Available at: www.scidev.net/global/features/the-rise-of-african-science-journalism-1/ (accessed 23 June 2021).

Newman, N., Fletcher, R., Kalogeropoulos, A. and Nielsen, R.K. (2019) *Reuters Institute Digital News Report 2019.* Available online at: https://reutersinstitute.politics.ox.ac.uk/sites/default/files/2019-06/DNR_2019_FINAL_0.pdf (accessed 30 December 2021).

Radford, T. (2008) Reporting on controversies in science. SciDev.Net, 13 February. Available at: www.scidev.net/global/practical-guides/reporting-on-controversies-in-science (accessed 23 June 2021).

Scott, C. (2012) Interview questions that work for newbie science reporters. International Journalists Network. Available at: www.aasciences.africa/aesa/programmes/science-communicationafrica-science-desk-asd (accessed 30 November 2021).

SPJ (Society of Professional Journalists) (2014) *SPJ Code of Ethics.* Available at: https://www.spj.org/pdf/spj-code-of-ethics.pdf (accessed 16 February 2022)

Waithera, H.W. (2018) *Science in Media: Africa Science Desk Baseline Assessment Report.* Available at: www.aasciences.africa/aesa/programmes/science-communicationafrica-science-desk-asd (accessed 23 June 2021).

Part II

Science Journalism in Practice

This part is designed to deliver hands-on skills for reporting science. It has ten chapters, each of which is based largely on the content of the Script programme's "Science Communication Skills for Journalists" online course. The content has been tried and tested by lecturers at four African universities, who used it to successfully train thousands of students in science communication prior to the publication of this book. Each chapter carries hands-on advice on the practice of science journalism, with learning activities to deepen your understanding of the topic.

Working with Scientists

In this chapter
- How to find and work with a credible scientist
- Why do you need a scientist anyway?
- How to identify a credible scientist
- Attribution
- Managing contacts
- Maintaining good relations
- Interviewing scientists: asking the right questions
- How to find a story at a scientific conference

There is no science story without scientists. Identifying the right scientists to interview, asking the right questions and making good use of scientific conferences are important tasks in science journalism.

How to find and work with a credible scientist

A surgeon's mistake happens behind closed doors and usually affects one patient; a journalist's error is public and can affect thousands of people. This is why it's essential to take care when choosing your sources of information.

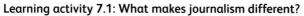

Learning activity 7.1: What makes journalism different?
In your view, what is the value that draws people to the news media even though the internet is overflowing with information on every topic? Write down one key word that represents the value added by the media.

Please write down your response before reading on.

Feedback

Verification.
    Audiences put their trust in the news media because the news media apply a process of verifying information. Every journalist has an ethical responsibility to protect that trust by, among other things, using credible sources.

©2022 CAB International. Science Communication Skills for Journalists: A Resource Book for Universities in Africa (Ed. Charles Wendo)
DOI: 10.1079/9781789249675.0007

Why do you need a scientist anyway?

Definitions of a scientist and of scientific work vary. For practical purposes, you can think of a scientist as, broadly, an expert who has knowledge on the topic you're writing about. This could be a scientific researcher or a professional who applies scientific knowledge.

You may need a scientist for the following reasons:

- to give you a tip-off about an interesting development in their field
- to give you a science angle on a running story
- to provide expert views and facts for a story
- to lead you to other human, paper or digital sources of information.

How to identify a credible scientist

Before choosing to interview a scientist, ask yourself: "Is this person qualified to give expert advice on the topic I'm reporting about?" Box 7.1 summarizes the criteria for selection of a scientist to interview, while the text below provides details on the criteria.

In choosing a scientist to interview, you can consider the following:

- **Their academic qualifications** – Institutions often have the profiles of their scientists on their websites, or in their annual reports and brochures. You can also find out about a particular scientist's qualifications through their professional bodies, by asking the scientist her/himself about them, or by checking their social media profiles. You could also note down any relevant information when a scientist introduces her/himself at a meeting.
- **Their area of specialization** – Even if they have high academic qualifications, one scientist will not be a suitable source of information for every topic – for instance, a professor of software engineering may not be the appropriate person to comment on malaria treatment, even if they have developed software for diagnosing malaria. Similarly, a professor of social sciences would be qualified to provide expert views on the socioeconomic aspects of genetic modification, but not the actual genetics.
- **The profile of their institution** – The profile of an institution influences the public's perception of a scientist. Furthermore, high-profile institutions are likely to have mechanisms that ensure their scientists are credible.
- **Their publications** – Good scientists publish; you can look up a scientist's name on Google Scholar to see what and how much they've published.

Box 7.1. Criteria for choosing a credible scientist.

- Academic qualifications
- Area of specialization
- Profile of their institution
- Their publications
- Whether their publications are cited by other scientists
- Presentations at scientific meetings
- Assignments with reputable organizations

- **Whether they are cited by other scientists** – A simple name search on Google Scholar can show you how often someone is cited by other scientists. If other scientists think this person is an authority on a given topic, then most likely they are.
- **Their presentations at scientific meetings** – Keep track of important scientific meetings in your country and region and get details of the speakers. Scientists who make key presentations at such meetings are likely to be highly respected experts.
- **Their assignments within national and international scientific bodies** – for instance, someone who is a member of a technical working group of a national or international body is likely be an expert in the subject concerned.

Learning activity 7.2: Most appropriate source

Tick the most appropriate source for a quote on the following subjects.
   A discussion on the usefulness and safety of an experimental Ebola vaccine

   a) A professor of anthropology who has studied human behaviour in previous Ebola outbreaks.
   b) A virologist who is attached to an infectious diseases institute.
   c) A veterinary epidemiologist who studies the transmission of Ebola from forest animals to human beings.

   How the construction of a dam is likely to affect the flow of water downstream

   a) A lecturer in hydrology at a national university.
   b) A climatologist working with an international institution.
   c) The head of the country's fisheries research institute.

Please write down your responses before reading on.

Feedback
   A discussion on the usefulness and safety of an experimental Ebola vaccine

   b) A virologist who is attached to an infectious diseases institute.

   How the construction of a dam is likely to affect the flow of water downstream

   a) A lecturer in hydrology at a national university.

Attribution

Using the right source is one thing, making sure your audiences know you are using the right source is another. Always attribute information to your sources.

How to find a credible scientist to use as a source

- **Journal articles usually indicate the official addresses of their authors** – You can use the address to find out about the author of a published paper you have read.
- **Referral from other sources** – If you write accurate, credible science stories your sources are likely to refer you to other credible scientists.
- **Meetings and events** – A journalist's achievement regarding a particular event is not the number of stories they write based on it, it is the number of contacts and story ideas they harvest from it. When attending events, introduce yourself to as many scientists as possible and exchange contact details with them, including those you do not immediately need.

- **Professional networks** – Some professional institutions and associations keep a list of relevant scientists.
- **Direct contact** – for instance, meeting a scientist at a conference or visiting their institution.

Managing contacts

A contact list is a journalist's goldmine and one of their most important assets. Having acquired the contact details of possible scientific sources, you need to ensure you can access them when you need them. Store your contacts in a format that is easily retrievable, accessible from any location and not vulnerable to theft. Digital tools, such as the Google Contacts app, Microsoft Outlook contacts function, Microsoft Access and a number of mobile apps are useful: they allow you to store someone's contact details and brief notes about them, and they can be backed up and synced so that if something goes wrong with your device you'll still have access to your contacts.

Learning activity 7.3: Contact lists

Look at this template for capturing the details of sources. The template is incomplete: can you identify the missing fields? List any additional fields/columns you would add.

Contact details for sources

Name	Institution	Email	Tel.	Website

Please write down your response before reading on.

Feedback
- Title (e.g. Dr, Mr, Professor)
- Area of specialization

Maintaining good relations

- **Break the ice** – Introduce yourself and establish a rapport with your contact. Scientists are more likely to talk to a journalist who has been in contact with them before.
- **Give them your contact details** – A scientist can tip you off about an important development, but only if they have your contact details.
- **Take their contact details and keep in touch with them** – After the initial contact, keep in touch with your source (but make sure you don't cross the line and impinge on their personal space).
- **Ensure accuracy** – Mis reporting a science story will put scientists off.
- **Don't misspell or get their name wrong** – Ask them for the correct spelling and pronunciation of their name.

- **Do some research about the subject** – Most scientists prefer to work with a journalist who has working knowledge of the subject concerned.
- **Make good when you make a mistake** – To err is human but mistakes should be followed by a correction.
- **Manage their expectations** – Don't promise something you can't deliver. For instance, don't say you'll let them read your copy before publication when you know you won't. Explain to them what's possible and what's not possible.

Learning activity 7.4: Should a scientist look at your story before you publish it?

Suppose at the end of an interview a scientist asks you to send them a copy of your story to verify the facts before publication. What would you tell them?

Please write down your response before reading on.

Feedback

This is one of the most contentious issues in science journalism. Most scientists prefer to be given the opportunity to check a story for accuracy before publication. Unfortunately, sometimes they seek to use this opportunity to revise their thoughts and censor your story. For most journalists, the answer to this request would be a straightforward "no" because it would compromise their journalistic independence. However, some journalists are happy to send back only those sections of the story that quote that particular scientist.

Whatever your thoughts, it is important to follow the guidelines of your media organization. Some media organizations allow their reporters to send a story to the scientist who provided information but the majority don't.

Interviewing scientists: asking the right questions

Interviewing is the most common way of getting information from a source. A story can only be as good as the interview. The basic rules of interviewing apply to science journalism as well. However, you need to bear in mind a few specific things about scientists when interviewing them.

Learning activity 7.5: Handling scientists during an interview

Below are a few of the challenges that journalists can face when interviewing scientists and reporting on science stories. Note down what you would do to overcome each of the challenges.

- Scientists are press-shy.
- Scientists use complicated language.
- Science is an unfamiliar topic and is difficult to understand.
- Scientific facts are delicate. A misrepresentation can be disastrous.
- A journalist may not know whether the research findings provided by the scientist are credible.
- It is often difficult to make scientific information relatable to the general public.

Continued

Learning activity 7.5. Continued.

Please write down your response before reading on.

Feedback

Challenge	What to do about it
Scientists are press-shy	Build a rapport beforehand: scientists are more likely to accept an interview request if you have interacted before.
Scientists use complicated language	If a scientist uses a word or concept you don't understand, ask them to explain it. It's better to ask repeatedly than to misrepresent scientific facts. Also ask the scientist to help you find alternative ways to explain the concept to your audiences.
Science is an unfamiliar topic and is difficult to understand	Do some background reading and research to get the basic information before you conduct the interview.
Scientific facts are delicate. A misrepresentation can be disastrous	Take notes carefully, record the whole interview and ensure all facts are presented accurately. At the end of the interview, restate the main points you have identified. When writing the story, don't hesitate to call back the scientist to ask for clarification whenever you need it.
A journalist may not know whether the research findings provided by the scientist are credible	Ask the scientist whether their findings have been published in a peer-reviewed journal.
It is often difficult to make scientific information relatable to the general public	Ask the scientist to point you to people who are affected positively or negatively by the issue under discussion.

Learning activity 7.6: Asking about research findings

Imagine you're going to interview a scientist about their research findings: list at least three important questions you would ask them.

Please write down your response before reading on.

Feedback

The following are the key questions that a story on research findings should answer.

- Who conducted the research?
- Where was the research done?
- When was the research done?
- When were the findings published?
- Which journal published the findings?
- What was the main research finding? What are the implications for society?
- What is new about this finding?
- What problem does the research finding solve?
- Is the finding conclusive, or does it need further research for confirmation?
- How do these results compare with other findings of similar research elsewhere?
- Who might these research findings affect (positively or negatively)?
- Should individuals, communities or the government change anything as a result of these research findings?
- Who financed this research?
- Has the research been peer reviewed and published in a journal?

From the questions suggested as responses to Activity 7.6, you can see that the 5Ws + H of journalism (who, what, when, where, why and how) apply to science reporting as well. These are generic questions that can help you plan an interview. In real life you will need to come up with more specific questions that suit a particular research study. You will be able to find the answers to some of these questions during your background research, while others are better provided by a scientist in their own words.

Learning activity 7.7: Questions to make the story relatable

Read the press release below and list three important questions you would ask the scientist(s) concerned, so as to make the story more relatable.

Fish consumption may prolong life

Consumption of fish and long-chain omega-3 fatty acids was associated with lower risks of early death in a *Journal of Internal Medicine* study. In the study of 240,729 men and 180,580 women who were followed for 16 years, 54,230 men and 30,882 women died. Higher fish and long-chain omega-3 fatty acid intakes were significantly associated with lower total mortality. Comparing the highest with lowest quintiles of fish intake, men had 9 % lower total mortality, 10 % lower cardiovascular disease mortality, 6 % lower cancer mortality, 20 % lower respiratory disease mortality, and 37 % lower chronic liver disease mortality, while women had 8 % lower total mortality, 10 % lower cardiovascular disease mortality, and 38 % lower Alzheimer's disease mortality.

Fried fish consumption was not related to mortality in men, whereas it was associated with increased risks of mortality from all causes, cardiovascular disease, and respiratory disease in women. Long-chain omega-3 fatty acid intake was associated with 15 % and 18 % lower cardiovascular disease mortality in men and women, respectively, when comparing the highest and lowest quintiles.

(Wiley, 2018)

Please write down your response before reading on.

Feedback

- We've always been told that fish is good for health: what is new about your findings?
- How should our diets change as a result of your findings?
- In real life, how much fish should we eat and how often?

Don't worry if some of your proposed questions are different from the suggestions above – you might just have chosen to focus on different concerns.

How to find a story at a scientific conference

Scientific conferences are an important source of stories, story ideas, scientists' contact details and resource materials for science journalists. However, to harvest these useful rewards requires a journalist's initiative.

Planning and creating a conference calendar

A credible scientific conference will publish its final programme months in advance. Study the programme and plan where to be on what day and at what time. Choose sessions according to the speaker or topic. The titles of presentations will give you a rough idea of where the story might be. Follow this up by reading the abstracts for the presentations (check the conference website for these). A conference abstract is similar to a journal abstract: the structure may vary but in most cases they begin with some form of introduction and end with the most important conclusions of the study.

Alternatively, you may decide to attend a session based on the name of the speaker, possibly because you anticipate an important revelation from that person, or you think they might make a statement that fits in well with a story you're working on. Don't choose a session simply because it has a high-profile scientist: not every high-profile scientist will have a scientific revelation to make at every conference. The biggest story of the conference might come from a lesser-known scientist who has a big research finding to reveal.

Finally, on session selection, don't follow the crowd. Some journalists make the mistake of "hunting in packs" at conferences: by congregating around famous persons they miss sessions that might have more important scientific findings. Also beware of activists – they are very effective at catching the attention of journalists but they could divert you from your original plan.

Having selected the session you're going to attend, use a calendar app or diary to help you organize yourself. Some conferences have an online programme where you simply click on the sessions you're interested in and it automatically generates a conference calendar for you. Allocate time to attending sessions and exhibitions, carrying out interviews and networking. Take note of the physical or virtual rooms hosting the sessions you're interested in, beforehand.

And finally, on preparation, you must consult or read about the current developments in the topic. In order to find out what's new, you need to know what's current. You could also look up the listed speakers and try to get an advance copy of their presentation. If they agree, make sure you observe the embargo and publish the story only after they present their paper.

At conference sessions

Whether it's a face-to-face conference or one held virtually, it's important to arrive or log in on time, to set up your recording equipment and to be settled before the presentation begins. It's important to listen to a presenter's introductory comments, because this will help you interpret the pronouncements that follow.

Whatever the circumstances, listening attentively and taking notes during the session is a basic requirement. You might have the most sophisticated recorder available but if you don't take notes you will have to waste a lot of time listening to the recording of the entire session in order to get the story. Taking notes while voice recording helps you to easily identify the key points and direct quotes.

During the discussions that follow a presentation, take note of what people say. Get their names so you will be able to quote them accurately.

At the same time, make sure you jot down any story ideas as they occur to you: don't trust your head to remember these ideas later. You can use notes apps, such as Simplenote, OneNote and Evernote, or even a notebook, to record your story ideas.

At exhibitions

Scientific conferences usually involve exhibitions (virtual or physical). An exhibition may appear to be a side activity at a conference but it can give you a big story. Take note of the items exhibited and ask questions about them where possible. Additionally, use the exhibition to network and expand your list of contacts. Exhibitions are also venues where you can find people to give you additional comments related to a story you've identified from a conference session.

Poster sessions

Not all papers at a conference are presented orally. Whether it's a physical or a virtual conference, some papers are presented in the form of a poster. Each poster is an abstract, usually interspersed with charts, graphs and pictures. The fact that a paper is presented as a poster and not in an oral session doesn't mean it isn't important. It's possible to find a big story in a poster session.

Further reporting

Because of time limitations, scientists' presentations may not answer all the questions you have in mind. After the session, you may want to ask the presenter a few questions. Make contact with, and seek an interview opportunity with, the scientist. You may also need the views of other scientists who were present in the same conference session: ask them what they make of the presenter's main pronouncement.

Finally, look for the missing voices. Ask yourself: "Who else might have something to say about this? Who else has a stake?" They could be other experts or people who have something to gain or lose from the new development. These could be affected persons, programme implementers or policymakers – and it's quite likely they won't be present at the conference.

Learning activity 7.8: Additional reporting

You are attending a scientific conference on HIV. A scientist presents her research findings indicating that she has detected some signs that some individuals might be resistant to HIV. Who of the following would be the most appropriate person to interview to gain a deeper understanding of the phenomenon?

a) An immunologist at a virus research institute
b) A person living with HIV
c) A minister of health
d) The head of the national drug authority

Please write down your response before reading on.

Answer
a) An immunologist at a virus research institute

Tips for finding a story

Here are four ways to utilize a conference to find a good story.

1. Listen to the presentations or speeches and pick out an interesting revelation.
2. Identify a story idea beforehand and come to the conference to collect information that feeds into it.
3. Carry out interviews or a survey during the conference and compile a story.
4. Make observations – for example, at the exhibition or demonstration – and pick out an interesting development.

Resource materials

You will most likely encounter a lot of resources at the conference, including website addresses, Facebook groups, mailing lists, Twitter accounts, books, magazines, brochures and flyers with information related to the theme of the conference. Collect as many of these as you can, and organize them for future use. Digital resources are easier to store than printed resources.

After the conference

- Sort the story ideas you've collected during the conference. Prioritize and create a timetable for handling them.
- Retrieve the list of contacts you met during the conference. Send each of them emails saying it was nice meeting them, and indicate how you would like to work with them going forward.
- Organize your informational material. You could bookmark the online resources or save the weblinks in your notes app.

Summary

- Every science journalist needs to build a list of credible scientists and to nurture professional relationships with them.
- Relying on inappropriate sources can mislead audiences and can dent the credibility of a journalist and their media house.
- In choosing a source for your story, consider their academic qualifications, their field of specialization, their institution, their publications, whether and how often they are cited by other scientists, their presentations at scientific meetings and their assignments within national and international bodies.
- Whereas basic interviewing skills are universal, there are some unique considerations you need to bear in mind when interviewing scientists, for example what is the main conclusion of your study and what are the implications for society?
- Adequate preparation will give you a great interview and a strong story.
- Effective prior planning will lead to more successful coverage of a conference.
- Before a conference opens, it is important to study the programme and create a personal calendar for covering it.
- Your success will come not only from the stories you write immediately after a conference but also from the story ideas and contacts you come out with.
- Don't underestimate exhibitions and poster sessions: they can give you great stories.
- Look out for information resources for future use.

Reference

Wiley (2018) Fish consumption may prolong life. *Eurekaalert.org*, 28 July. Available at https://www.eurekalert.org/pub_releases/2018-07/w-fcm071718.php#:~:text=Consumption % 20of % 20fish % 20and % 20 long,men % 20and % 2030 % 2C882 % 20women % 20died (accessed 28 April 2021).

Getting a Story From an Original Research Paper

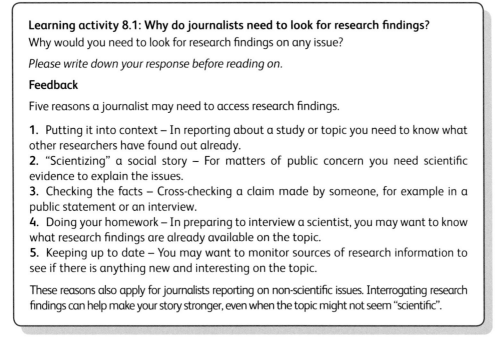

In this chapter
- Why every journalist needs to access research papers
- Using databases to find research papers
- How a research paper is structured

Thousands of scientific journals are published globally every year, and between them they present millions of research papers. A good number of research papers will be newsworthy. Considering the number of journals and research papers published annually, a curious journalist who gains access to scientific journals can never lack a research-based story.

Why every journalist needs to access research papers

As journalists, we often get our science stories from press releases and then look for human sources to strengthen our reporting. However, it is also useful to access research findings by looking at the original paper.

Learning activity 8.1: Why do journalists need to look for research findings?
Why would you need to look for research findings on any issue?

Please write down your response before reading on.

Feedback

Five reasons a journalist may need to access research findings.

1. Putting it into context – In reporting about a study or topic you need to know what other researchers have found out already.
2. "Scientizing" a social story – For matters of public concern you need scientific evidence to explain the issues.
3. Checking the facts – Cross-checking a claim made by someone, for example in a public statement or an interview.
4. Doing your homework – In preparing to interview a scientist, you may want to know what research findings are already available on the topic.
5. Keeping up to date – You may want to monitor sources of research information to see if there is anything new and interesting on the topic.

These reasons also apply for journalists reporting on non-scientific issues. Interrogating research findings can help make your story stronger, even when the topic might not seem "scientific".

©2022 CAB International. Science Communication Skills for Journalists: A Resource Book for Universities in Africa (Ed. Charles Wendo)
DOI: 10.1079/9781789249675.0008

Learning activity 8.2: Why do journalists need to look for research findings?

It is useful to make use of research findings, even for non-scientific topics. With this in mind, read the scenario below and answer the questions that follow.

Members of parliament in your country are complaining that there aren't enough female teachers to provide teaching cover in girls-only schools. They claim the shortage of female teachers compromises the wellbeing and academic performance of girls. Therefore, they propose a government policy to increase the number of female teachers through affirmative action. They are mobilizing other legislators to support their policy proposal.

Your editor has asked you to report on the story. You have decided that as well as interviewing human sources you will look at relevant scientific evidence for the legislators' claims.

1. What are the key assumptions the legislators may be making?
2. What type of scientific research might be useful to check the legislators' claims?

Please write down your response before reading on.

Feedback

Assumptions the legislators might be making

- Students in girls-only schools learn better when taught by female teachers.
- Students in girls-only schools have a better sense of wellbeing when taught by female teachers.
- Increasing the number of female teachers in girls-only schools will improve the academic performance and wellbeing of the students.
- There aren't enough female teachers for the current number of all-girls schools.

Research that might be useful to check the legislators' claims

- Performance data on school pupils, comparing boys and girls.
- Scientific research evidence on why boys might perform better than girls or vice versa.
- Studies investigating whether girls perform better when taught by female teachers.
- Data on the number of male and female teachers currently in schools.

Using databases to find research papers

Sources don't always provide evidence for the claims they make during interviews with journalists. Even if they do, their evidence could be biased or partial. Finding scientific evidence independently can help you check the claims.

Learning activity 8.3: Finding the right evidence

Where would you get the scientific information to help you make sense of the legislators' claims in the above scenario?

a) Wikipedia
b) A Google search
c) Research databases

Please write down your response before reading on.

Answer

c) Research databases

Wikipedia will help you get an initial handle on a topic, but be careful: anyone can edit an entry, the entry could be a work in progress, or it could be simply wrong. Entries are usually verified by the community, but there is no guarantee. Always check the references and the sources used in a Wikipedia entry.

Google has become the journalist's best friend. We tend to use a Google search for everything, but an ordinary Google search is not the right way to identify scientific evidence. It's good for general information, to give you an idea about a topic, and it might also give you some clues about the type of information that is out there, but some of the information thrown up by a Google search may be unscholarly, unverified or even false. This means you may not be able to use it to back up an argument.

With research databases, you know the information is verified. Databases are also often curated, organized, and described in useful ways, unlike the results of a Google search or Wikipedia entries. This can make searching easier and more powerful. Some databases allow you to create an account and to keep track of your searches, and to build up a library of sources.

Search techniques

To find information online you first need to ask questions and then try to find the answers to those questions. For example, you might pose the question: "Does climate change increase the spread of malaria in Africa?" To find useful research on this issue online, you would need to identify the keywords and concepts that make up the question. In this case the key terms and concepts could be "climate change", "malaria" and "Africa".

Learning activity 8.4: Key concepts of research

In the scenario above about female teachers (Activity 8.2), what are the key terms or concepts?

Please write down your response before reading on.

Feedback
- All-girls schools.
- The impact of female teachers on pupils.
- Factors affecting pupils' wellbeing and performance.
- The male to female teacher ratio.
- The sex gap in attainment.
- Schoolgirls.

Using alternative words and phrases in your search

To obtain better search results, you need to think of any alternative words and phrases that refer to the same thing. By broadening your search terms, you will avoid missing out on useful sources.

The following learning activities look at some of these searching techniques. You need internet access to carry out these activities. (To undertake these activities you will be asked to use an ordinary Google search for the search techniques, but these techniques also apply to searches of research databases.)

Learning activity 8.5: Alternative words and phrases

List possible alternative words or phrases for the following:

- All-girls schools.
- The impact of female teachers on pupils.
- Factors affecting pupils' wellbeing and performance.
- The male to female teacher ratio.
- The sex gap in attainment.
- Schoolgirls.

Please write down your response before reading on.

Feedback

Term or concept	Alternative words or phrases
All-girls schools	Girls-only schools, girls' schools
The impact of female teachers on pupils	Women teachers, effect on outcomes, teacher impact disaggregated by sex
Factors affecting pupils' wellbeing and performance	Welfare, comfort, health, socioeconomic factors, academic achievement, learning outcomes, effectiveness, intervention types, learning performance predictors
The male to female teacher ratio	Ratio of male to female teachers, percentage of female teachers, female teacher ratio, male vs female teachers
The sex gap in attainment	Performance differences, outcomes, disadvantaged students
Schoolgirls	Students, pupils, learners

Learning activity 8.6: Searching for phrases

1. Perform a Google search for "girls' schools" but don't use quotation marks around the search terms. Check the results: are the words "girls" and "school" always together?

2. Now perform a Google search for "girls' schools", using quotation marks. What do you notice about the results?

Please write down your response before reading on.

Feedback

The second search returns results where the two words are together. This is more precise in the context of the research we are interested in. The first search returns many pages we are not interested in. (Note: most pages with a search functionality allow you to do this.)

Learning activity 8.7: Using alternative words and phrases

Sometimes you will want to widen your search because there might be more than one way to describe the same thing.

1. Perform a Google search for "all-girls schools" + "female teachers". How many results has Google returned? Note this down.

2. Now search for "all-girls schools" OR "girls-only schools" + "female teachers". How many results has Google returned? Note this down.

Continued

Learning activity 8.7. Continued.

What difference do you notice between the first and second searches?

Please write down your response before reading on.

Feedback

The second search should have returned more results. You can make your search more inclusive by adding the word "OR" in capital letters and then adding a synonym for your search term.

Learning activity 8.8: Using truncation

Some words can have varying forms: for example, differences at the end of a word depending on whether the British or American spelling is used (e.g. Neigh*bour* versus Neigh*bor*). The root of a word can also be part of many other related words having to do with the same topic. Truncation can help you capture these.

1. Go to the free research database Pubmed Central (https://www.ncbi.nlm.nih.gov/pmc/journals/) and search for "climate". How many results has the search returned? Write this down.
2. Next, search "climat*" using the same database. How many results has the search returned? Compare this to your first search.

What difference do you notice between the results of the first and second searches?

Please write down your response before reading on.

Feedback

The second search should have returned more results. You can make your search more inclusive by using an asterisk to capture variations of a word. Using climat* will capture *climate, climatic* and *climatesmart*. This is called **truncation**. This technique may not work in a Google search, but it will work when searching most research databases.

Google vs Google Scholar

Google offers an academic search engine called Google Scholar.

Learning activity 8.9: Comparing Google to Google Scholar

1. Perform a search using the terms "female teachers" + "girls schools" in Google. Take note of the search results.
2. On a different tab, go to Google Scholar (https://scholar.google.com/) and enter the same search terms.

What differences do you observe between the results of the two searches? Look out for:

- the number of search results
- the type of research in the results
- the type of documents in the results.

Please write down your response before reading on.

Continued

Learning activity 8.9. Continued.

Feedback

The ordinary Google search returns more results than Google Scholar. Google Scholar is more focused on scholarly papers while the ordinary Google search picks up anything related to the topic. With Google Scholar you can also navigate existing literature on a topic by clicking on citations and related articles.

Google Scholar vs research databases

It might be tempting to think that you can just use Google Scholar to find research. However, Google Scholar doesn't have the same level of curation as a research database. When looking for scientific research findings, Google Scholar is more reliable than ordinary Google searches, but searching specialized databases is even better, for several reasons:

- information in research databases has been verified and indexed
- some research databases have digital copies of hard copy journal articles – you may not find these through a Google Scholar search
- Google Scholar cannot find all the articles contained in a research database
- some information is not free online – a Google Scholar search may lead you to the title but when you click on it the system may ask you to pay for access.

Subscription vs free databases

Some research databases are free while others are paid for. Here are some examples of free research databases:

- Pubmed Central (https://www.ncbi.nlm.nih.gov/pmc/) – A free digital repository that archives open access full-text scholarly articles that have been published in biomedical and life sciences journals.
- Open Science Directory (https://opensciencedirectory.net/) – A search tool that can be used to search open access journals.
- Directory of Open Access Journals (https://doaj.org/)– A community-curated online directory that indexes and provides access to high-quality, open access, peer-reviewed journals.
- Open Directory Project (http://odp.org/) – An open content directory that provides links across a range of subjects.
- Arxiv (https://arxiv.org/) – An open access repository of electronic preprints approved for posting after moderation (but not peer review).
- African Journals Online (https://www.ajol.info/) – The world's largest platform of African-published scholarly journals.
- AIDSInfo (https://aidsinfo.nih.gov/) – A US Department of Health and Human Services website for federally approved HIV/AIDS medical practice treatment guidelines.

Here are some examples of paid-for databases:

- CAB Direct (https://www.cabdirect.org/) – A thorough and extensive source of references in the applied life sciences, incorporating the leading bibliographic databases *CABI Abstracts* and *Global Health*.
- Science Direct (https://www.sciencedirect.com/) – A website which provides access to a large bibliographic database of scientific and medical publications of the Dutch publisher Elsevier.
- JSTOR (https://www.jstor.org/) – A digital library of academic journals, books and primary sources.

How a research paper is structured

The writing styles of journalists and of scientists differ considerably. While both identify a problem, ask questions, look for the answers and write up their findings, journalists usually place the news at the top of their articles, while scientists "hide" it at the bottom of theirs. For this reason, when reading the first few paragraphs of a research paper it can be easy for a journalist to incorrectly conclude that it has no news. Knowing the structure of a research paper can help you to identify the story easily, without being intimidated by scientific content. Most research papers follow the structure presented in Fig. 8.1. On the contrary, most news stories about scientific research are written in the form shown in Fig. 8.2.

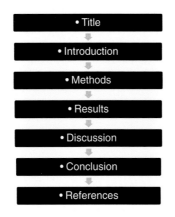

Fig. 8.1. Typical structure of a research paper.

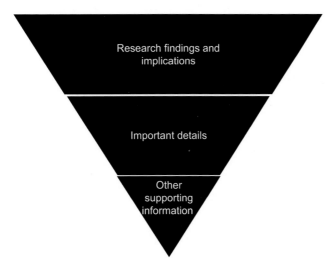

Fig. 8.2. Typical structure of a news story about scientific research.

Now, let's look at the structure of a scientific paper in more detail, and how you can use the different sections of a research paper to develop your story.

Understanding the title

Consider the research paper titled "Perioperative patient outcomes in the African Surgical Outcomes Study: a 7-day prospective observational cohort study" (Biccard *et al.*, 2018). From this paper, a journalist reported a science news article with the headline "African patients more likely to die from surgery" (Otieno, 2018).

Let's look at how a journalist might derive a science story from this research paper. The title of the research paper doesn't reveal the research findings, but certain keywords ("patient", "surgical" and "outcomes") give you a clue that the study could be about the fate of patients who undergo surgery.

If you are not sure about your interpretation of the title or any fact in the research paper, ask a scientist.

Some questions might come to mind when you look at the title of this research paper. Do the patients come out of theatre alive? Do they recover from their condition? Do they come out of the operation with some damage? What happens to them afterwards?

Learning activity 8.10: Interpreting titles

Below are two titles from research papers.

Title 1: "Associations of fats and carbohydrate intake with cardiovascular disease and mortality in 18 countries from five continents (PURE): a prospective cohort study" (Dehghan *et al.*, 2017)

Title 2: "Antimicrobial Resistance: The Major Contribution of Poor Governance and Corruption to This Growing Problem" (Collignon *et al.*, 2015)

Try to interpret these titles and think of the type of story you could write based on each one. Use the following questions/tasks to help you:

1. What is each of these articles about?
2. Choose one of them and suggest interesting questions for a news article based on them.

Please write down your response before reading on.

Feedback

Here is a simpler alternative to each title and some interesting questions about each one.

Title 1: "The link between heart disease and the consumption of fats and carbohydrates"

- Which foods contain these fats and carbohydrates?
- Which countries was the research carried out in?
- How many people are affected?

Title 2: "How poor governance and corruption contribute to drug resistance"

- Which countries is this a problem in?
- How many people are affected as a result?
- Which microbes (and therefore diseases) are thriving as a result?

Finding the story in a research paper's abstract

Before reading a full research paper, scan the **abstract** to see if there could be a story there. An abstract is usually a 200–300 word summary of the paper, which should take two to three minutes to read. In a few words, it helps you to know why the study was necessary, how it was done, its main findings and its conclusions.

The **conclusion** part of the abstract, which some researchers refer to as the "interpretation" or "implication", often reveals the story. In our surgery example above, the interpretation in the abstract reads: "Despite a low-risk profile and few postoperative complications, patients in Africa were twice as likely to die after surgery when compared with the global average for postoperative deaths."

Certainly, there is a story here. However, before we proceed with it, if we have any doubts about our interpretation of the statement we must check it with a scientist.

Having considered the abstract, you should read the rest of the paper to get further insights. The **introduction** will give you a lot of useful background. The **methodology** section will describe how the research was done. The **results** section of the research paper will explain the findings. In the **conclusion**, the researchers will summarize their interpretation of the research and its implications.

Beyond the research paper

Remember that a research paper is only a starting point. For a great story, you will need to interview the researcher and a number of other sources. Sourcing is discussed in another chapter.

Summary

- A science journalist often needs access to research publications for fresh stories or to get scientific evidence to enrich a story.
- An ordinary Google search and a Wikipedia entry may be good for general information but cannot be relied upon for scientific evidence.
- Research databases provide verified information that has been scrutinized by professionals.
- Google Scholar is more credible than an ordinary Google search, but is inferior to research databases.
- Identify the concepts that make up a question in order to return better results.
- Use operators such as AND and OR to link your concepts when searching.
- The table below summarizes what you can use each tool for.

Tool	Use
Google search	For general information, but may not be relied upon for scientific evidence
Google Scholar	To complement research databases when searching for academic literature
Research databases	For finding studies, because the search is more focused and the results are more reliable

- The title of a research paper will give you a clue and the conclusion part of the abstract will unveil the story.
- You don't have to know all the scientific terms in a research paper to identify a story: keywords can give you an indication of where the story is.
- A research paper is only a starting point: to get a great story you will need multiple sources.

References

Biccard, B.M., Madiba, T.E., Kluyts, H-L., Munlemvo, D.M., Madzimbabuto, F.D., Basenero, A., Gordon, C.S., *et al.* (2018) Perioperative patient outcomes in the African Surgical Outcomes Study: a 7-day prospective observational cohort study. *Lancet* 391(10130), 1589–1598.

Collignon, P., Athukorala, P-C., Senanayake, S. and Khan, F. (2015) Antimicrobial resistance: The major contribution of poor governance and corruption to this growing problem. *PLOS ONE* (online) 18 March. Available at: https://journals.plos.org/plosone/article?id=10.1371/journal.pone.0116746 (accessed 29 April 2021).

Dehghan, M., Mente, A., Zhang, X., Swaminathan, S., Li, W., Mohan, V., Iqbal, R., *et al.* (2017) Associations of fats and carbohydrate intake with cardiovascular disease and mortality in 18 countries from five continents (PURE): a prospective cohort study. *The Lancet* (online) 29 August. Available at: https://www.thelancet.com/action/showPdf?pii=S0140-6736 % 2817 % 2932252-3 (accessed 29 April 2021).

Otieno, S. (2018) African patients more likely to die from surgery. SciDev.net, 22 January. Available at: http://www.scidev.net/global/news/african-patients-more-likely-to-die-from-surgery-1x/ (accessed 29 April 2021).

Getting a Science Story From Technical Reports

9

In this chapter
- Technical reports as a source of news
- Types of technical report
- How technical reports are structured
- Getting a story from statistical tables
- How to move from numbers to an interesting and informative story

---

Technical reports as a source of news

Technical reports published by local and international organizations often provide information that may be of interest to the wider public. In this regard they can be a great source of stories for journalists: they can provide a fresh story, a new angle to an old story or useful information that can add context to a story you are working on. Organizations publish reports to communicate facts, trends and their positions on a topic or a sector.

Learning activity 9.1: Technical reports vs news

List the reasons institutions publish technical reports. Separately, list why journalists report news. What do you see in common between why journalists report news and why institutions publish technical reports?

Please write down your response before reading on.

Feedback

Both technical reports and media stories are published to disseminate information, though media stories have a wider audience and simpler language. It's not surprising therefore that some of the information in a technical report may be newsworthy.

Types of technical report

Often technical reports from scientific institutions provide scientific information that should be shared beyond the scientific community. As illustrated in Fig. 9.1, there are many types of technical report, for example:

©2022 CAB International. Science Communication Skills for Journalists: A Resource Book for Universities in Africa (Ed. Charles Wendo)
DOI: 10.1079/9781789249675.0009

Fig. 9.1. Types of technical report.

- background reports – which provide contextual information on an issue
- progress reports – which show how far the organization has moved towards the completion of a project
- periodic reports – which provide a regular summary of activities and achievements during a specific period, for example monthly, quarterly or annually
- trends reports – which show changes in key indicators, such as demographics, over time
- feasibility reports – which project the workability of a planned project
- evaluation reports – which provide an assessment or analysis of a current project, or a judgement/analysis on how well a project is performing.

All of these reports provide technical information that may be of interest to the wider public – and they usually do so using more reader-friendly language than is used in scientific papers that are meant for publication in a peer-reviewed journal.

How technical reports are structured

Reports vary in structure but most of them will have a cover page, a title page, an executive summary, acknowledgements, a table of contents, an introduction, the main body of the report, conclusions, references and annexes. The main body of the report may be broken down into chapters.

The cover page
If you're on the lookout for a newsworthy story and are approaching a technical report to see if it contains one, you need to examine every component of the cover page to understand what information the report contains and to judge how useful it may be to you.

Learning activity 9.2: The cover page

Take a look at the cover page of the UN Food and Agriculture Organization's (FAO's) 2018 report *The State of the World's Forests*.

Source: FAO (2018a)

Continued

Learning activity 9.2: Continued.

Now answer the following questions:

1. What is the name of the organization releasing the report?
2. Assuming you're reading this in July 2018, would you consider the report current: yes or no?
3. Bearing in mind the title, which of the following topics would you **NOT** expect to find covered in the report?
   a) Forest ownership
   b) Agriculture area with tree cover
   c) Cattle population per square kilometre
   d) Percentage of land covered by forests

Please write down your response before reading on.

Feedback

1. The FAO.
2. Yes.
3. c) Cattle population per square kilometre.

The cover page bears the title of the report, the name of the authoring institution and the date of publication. The title gives you an idea what the report is all about and what kind of story you might find in it. The name and logo of the organization issuing the report will determine whether you can rely on it or not: it's essential to check the credibility of the authoring organization before using any report as the basis for a story. Names and logos of partner institutions give you an idea who else is involved. The date of publication tells you something about the freshness of the information – you wouldn't want to write a news story derived from a report that is two years old. However, an old report can give you some useful background information to use in your story.

Table of contents

The table of contents is a comprehensive listing of what is in the report, section by section. It can be a useful pointer to where you might find the information you need. Just browse through the table of contents quickly and proceed to the executive summary. You will need to come back to the contents after reading the executive summary, to identify the sections of interest to you.

Executive summary

The executive summary highlights the key points that are presented in the report. It will most likely take you less than five minutes to read. Normally each section of the report will be summarized in the executive summary. From the executive summary you can select an area of interest, then use the table of contents to find the relevant section in the main body of the report in order to get the details. Some reports provide a few sentences or bullet-point summaries at the beginning of each section or chapter, summing up the key information, instead of a full executive summary at the beginning of the report. In that case it is advisable to skim through the summary of each section to have a rough idea where the story might be, before reading the details.

Learning activity 9.3: Seeing the story in an executive summary

Take a look at the one-page executive summary below and pick out at least one fact that catches your attention and that you think merits further investigation for a possible story. Write this fact down in a sentence of not more than 20 words.

Executive summary

Background

Invasive species are estimated to cost the global economy over US$1.4 trillion annually and have negative impacts on the livelihoods of vulnerable communities, driving food insecurity and undermining ongoing investments in development. The geographic spread and impact of invasive species is accelerating due to climate change, trade and tourism.

CABI's new invasive species programme

Since 1927, CABI has led or made major contributions to over 50 invasive species management projects across Africa, Asia and the Americas that used biocontrol as a key component of Integrated Pest Management (IPM). During the last round of consultations (2015–2016) with its 48 member countries, CABI received unanimous endorsement of the need for systematic action, with widespread recognition of the threat of invasive species and the value that a comprehensive CABI-led programme would bring, CABI has therefore repositioned its approach to combat the threat and impact of invasive species by building a cross-sectoral coalition of interested stakeholders to operate a three-stage approach to the problem: prevention, early detection and rapid response, and control and restoration. The programme will build on CABI's global experiences in managing invasive species to improve rural livelihoods, food security and trade, as well as protecting agricultural and natural ecosystems.

Progress in 2014–2016

Pre-implementation activities for the invasive species programme were carried out between 2014 and 2016, with successes in three areas:

Building the evidence base on invasive species and their impacts: the impacts of invasive species on the economy and rural livelihoods have been investigated, consultations have taken place with international researchers on the link between invasive species and malaria, and methods for managing invasive species have been investigated. As a result of this work, a number of scientific journal articles and one book chapter have been published.

Raising awareness of the problem of invasive species: a website on **invasive species** has been created, the invasive species programme has been promoted at international events (eg COP13, EcoSummit), CABI has worked with **SciDev.Net** on an awareness campaign on the problem, and management guides for certain species have been developed and distributed.

Preparing for implementation of the invasive species programme: consultations have taken place with CABI member countries, distribution maps of invasive species that pose a threat have been developed, and key documents for programme implementation have been drafted (including the programme strategy and theory of change, and technical documents setting out CABI's approach to determine which invasive species can and should be prioritised within a country and region).

Next steps

The next steps will build on this strong foundation to implement the programme. Under the current funding for the programme, planned activities are ongoing – principally in Ghana and Pakistan, but also at an international level – to generate partnerships and policy awareness; build capacity in the areas of prevention, detection and response and control; reach and advise rural households about biological invasions; and develop invasive species information resources and tools.

At the same time, CABI is seeking a US$50 million investment to implement the invasive species programme globally, with the aim to protect and improve the livelihoods of 50 million poor rural households impacted by the worst invasive species. This money will be spent on four main activities: collecting and sharing of data and knowledge; fostering partnerships at regional, national and local levels; validating management options and building capacity; carrying out community-level activities at scale. Cross-cutting work will cover monitoring and evaluation, and gender empowerment and youth inclusion.

Required funding

The invasive species programme strategy has been endorsed by CABI's 48 member countries and CABI is now ready to implement the programme. The proposed investment of US$50 million in the programme will help to reverse the invasive species threat and protect 50 million rural families from devastating losses.

Source: CABI (2017)

Please write down your response before reading on.

Feedback

The following are potentially newsworthy facts from the executive summary.

1. Invasive species cause economic losses worth over US$1.4 trillion per year globally.
2. Climate change and tourism have speeded up the spread of invasive species.
3. Scientists have created a distribution map showing the hotspots for invasive pest species.

Main body of the report

The main body of a report usually provides a comprehensive presentation and discussion of the key issues. By the time you get to the main body of the report you will have a rough idea where the story is and what the issues might be. As you read, bear in mind what could make a news item: for example, are there any surprising revelations, policy changes or dramatic numbers? Whenever you find a newsworthy piece of information, underline it or write it down, indicating the page number.

Take your time to read and understand so that you can report accurately. It is important to take notes to guide you when writing the story. As you read the report, certain questions will come to mind: note them down so that you can ask the relevant persons afterwards. The report may have some stand-alone content – for example, short success stories, case studies and fact boxes – pay attention to these too.

Reports often have statistical tables in the main body and in the annexes. Later in this chapter we will look at how to find the story in a statistical table.

Annexes

Annexes usually include details or supporting information that is not essential and so is not placed in the main body of the report. They can enable you to get additional information as well as more information about the context of the report. For example, if the report is based on a survey, the questionnaire that was used to conduct it may be contained in an annex. You will want to see this so you know precisely how a question was asked, which will enable you to make a better interpretation. For example, the question "How often do you fumigate your house?" will generate a different answer to "How many times have you fumigated your house in the last 12 months?".

Beyond the report

Remember that a report is only a starting point for your story. For a good story you will need to interview a number of people. Capture all the voices that need to be expressed in the story. Map the stakeholders and speak to them. Who are the key players? Who are the affected persons? Who else would justifiably have something to say about this?

Don't forget the basic principles. A good science story has the following attributes:

1. It is comprehensible.
2. It is relatable.
3. It has credible sources.
4. It is accurate.

The CNN story: "Can the Middle East solve its water problem?" (see https://edition.cnn.com/2018/07/11/middleeast/middle-east-water/index.html) (Scott, 2019) is a good example of how a journalist can produce an interesting and informative story for the mainstream media, based on a technical report. The story is based on a World Bank report, *Beyond Scarcity: Water Security in the Middle East and North Africa* (see https://openknowledge.worldbank.org/handle/10986/27659) (World Bank, 2018).

Getting a story from statistical tables

Technical reports often include lots of statistical tables. Studying these tables is a good way to quickly spot the story in a technical report.

Learning activity 9.4: Find the story on this page

Take a look at the page below, extracted from the FAO's field guide on management of the Fall Armyworm, and identify and write down at least one piece of information that is potentially newsworthy that you can report about.

Don't panic! Maize plants can compensate significant damage by the Fall Armyworm

Damaged plants can scare farmers. Never before have they seen this type of damage, where the insect eats through so much of the leaves. Farmers know about stem borers, but because they aren't often seen (hidden in the stems), they don't often scare farmers like this new pest, Fall Armyworm.

The spectacular-looking damage is very photogenic. The combination of farmers' nervousness, media alarmism, and politicians' quick reaction to do something has led to some bad decisions, including the use of Highly Hazardous Pesticides. Some older pesticides, which have long been banned in other parts of the world due to demonstrated human health impacts, are still available and used in some African countries. Some of the older pesticides don't work, because FAW has developed resistance to them[1].

Such panicky responses are likely when the farmers and others don't understand the potential impact of FAW damage. The quick response to sight of significant-looking damage is to assume that it will cause dramatic yield reduction. But that's not necessarily true. In fact, we know that in most cases FAW does NOT cause "total destruction". In most cases the leaf damage does cause some yield reduction, but it is probably far less than what farmers without experience with the pest believe.

Maize has been selected by humans for thousands of years to yield well, even in face of damage to insects, pathogens and other threats. These eons of selection have resulted in maize plants that have considerable capacity to compensate for foliar damage.

The response of maize yield to FAW infestation has been studied in the field a number of times in the Americas. A review of these studies shows that while of concern, FAW damage in maize is not devastating. While a few of the studies show yield reductions due to FAW of over 50 percent, the majority of the field trials show yield reductions of less than 20 percent, even with high FAW infestation (up to 100 percent plants infested). Maize plants are able to compensate for foliar damage, especially if there is good plant nutrition and moisture. While FAW needs to be managed sustainably by farmers, it is not cause for panic.

In FFS, we can examine our maize's ability to compensate for defoliation by conducting a Special Topics experiment (see section B.6.7). The experiment will look at the impact of defoliation of maize plants at different growth stages on grain yield.

1. For FAO guidance on which pesticides used on FAW in Africa might be Highly Hazardous Pesticides, see FAO Guidance note on Reducing risk from pesticides at: http://www.fao.org/food-chain-crisis/how-we-work/plant-protection/fallarmyworm/en

../cont.

Source: FAO (2018b)

Please write down your response before reading on.

Feedback

Two examples of newsworthy information are:
- new evidence shows that maize plants can recover from the damage caused by the Fall Armyworm if the plant is well-nourished
- when the Fall Armyworm attacks, some farmers panic and needlessly spray their maize with highly hazardous pesticides.

Why the numbers are important

In March 2018 SciDev.Net's capacity-building programme, Script, held a three-hour training and networking workshop in Kigali, Rwanda for scientists and journalists. Each journalist was given an opportunity to speak to several scientists and ask them about their research. At the end of the session, scientists were asked what they had learnt from the exercise. One scientist said: "Now I know what journalists want. Most of them were asking for statistics – how many this, how much that …"

Does that sound familiar? When you tell your editor about a disease outbreak, they will immediately ask you how many people have died or become ill. When a local leader tells you about a food shortage in their area, you will immediately ask them how many families are starving or which communities are most affected. Each of these cases involves numbers and statistics, which are often provided in the form of tables.

Statistical tables are regular features in technical reports and research papers, but most journalists would rather avoid them, largely because they're perceived to be difficult to understand. If you are among the majority that get intimidated by statistical tables, don't worry: we will take you through a step-by-step process for easily getting a good story from a statistical table. A journalist who knows how to read such tables will get more and better stories from a report.

Learning activity 9.5: Necessary voices

Assume you are working on a story about a technical report that reveals the economic damage caused by a particular invasive pest species. What are the categories of people you would consider interviewing for your story, and what information would you need from each category?

Please write down your response before reading on.

Feedback
- Farmers, for a real-life experience.
- Agricultural government officials, for what the government is doing about it.
- An expert from a university, for an independent scientific opinion.
- The authors of the report, for any clarifications or additional information.

Learning activity 9.6: Why use numbers and statistics in stories?

List some of the reasons why it's important for a journalist to report numbers and statistics.

Please write down your response before reading on.

Feedback
- Numbers and statistics help us to understand and explain the significance of what we are reporting about.
- Numbers and statistics help us to measure any progress (or lack thereof).
- Numbers and statistics help us to make comparisons between people, places and situations.
- Therefore, numbers and statistics help us to better understand the world we live in.

Different ways we use statistics

Statistics can be dramatic enough to make news straight away. For example, this Voice of America story: "42 percent of Americans say teen sex is morally acceptable" (see https://learningenglish. voanews.com/a/percent-of-americans-say-teen-sex-morally-acceptable-/4476713.html) (VOA News, 2018).

In other situations, you may have a story that is based on an event or an observation and you will want to substantiate it by adding statistics. An example is this story: "Vasectomies slowly gaining acceptance in Kenya" (see https://www.voanews.com/a/vasectomies-slowly-gaining-acceptance-in-kenya/4413227.html) (Ombuor, 2018).

Understanding tables

To get a story from a statistical table, it is important to know how a table is structured. Tables have **rows** and **columns**. Rows appear in a left to right orientation while columns appear in a top to bottom orientation. The **title** tells you what the table is all about. The **column header** is a word or phrase that tells you what information is in a given column. A **row header** is a word or phrase that tells you what information is in a particular row. Every row or column is divided into **cells**.

Number of boys and girls in Class X

	Boys	Girls
2018	26	14
2019	28	12
2020	24	16
2021	21	19

The **title** of the table above is "Number of boys and girls in Class X". The column headers are "Boys" and "Girls". Every column header tells you what is in that column: for example, every figure under the column that has the header "Girls" refers to the number of girls. The row headers are "2018", "2019", "2020" and "2021". Every row header tells you what is in that row: for example, every figure in the row that has the header "2020" represents the number of students in that year. Every cell is identified by the corresponding row and column headers: for example, the shaded cell with the figure "24" is the number of boys in Class X in 2020. Likewise, the shaded cell with the figure "14" represents the number of girls in Class X in 2018.

Finding the story

Here are a few points to look out for in a statistical table, which might indicate something worth investigating for a story:

- Do some columns have visibly higher or lower numbers than others?
- Do some rows have visibly higher or lower numbers than others?
- Within a row or a column, are there particular cells that stand out because they have numbers that are either far above or far below the others?
- Are there particular cells with figures that surprise, impress, excite, alarm or catch your attention in any other way?

Learning activity 9.7: Internet users in Africa

Take a look at the table below, which has been extracted from the *UNESCO Science Report 2010* (see https://unesdoc.unesco.org/ark:/48223/pf0000189958_eng), and answer the following questions:

1. Which region had the lowest internet use in 2008?
2. Which region had the fastest growth in internet use between 2002 and 2008?

Table 5: Internet users per 100 population, 2002 and 2018 (UNESCO, 2010)

	2002	2008
World	10.77	23.69
Americas	27.68	45.50
Europe	24.95	52.59
Africa	1.20	8.14
Asia	5.79	16.41

Please write down your response before reading on.

Answers

1. Lowest internet use: Africa.
2. Fastest internet growth: Africa.

Learning activity 9.8: HIV statistics

Now take a look at another table below, which is based on figures provided in the UNAIDS *Data 2017* report (UNAIDS, 2017).

Regional scorecard on UNAIDS targets

Region	Percentage of people living with HIV who know their HIV status	Percentage of people living with HIV who know their status who are on antiretroviral treatment	Percentage (of those taking antiviral medication) whose viral load has been suppressed
Eastern and southern Africa	76	79	83
Western and central Africa	42	83	73
Asia and the Pacific	71	66	83
Latin America	81	72	79

1. List at least three row headers in the table.
2. List all the column headers in the table.

Continued

Learning activity 9.8: Continued.

Please write down your response before reading on.

Answers

Row headers:

- Eastern and southern Africa
- Western and central Africa
- Asia and the Pacific
- Latin America

Column headers:

- Percentage of people living with HIV who know their HIV status
- Percentage of people living with HIV who know their status who are on antiretroviral treatment
- Percentage (of those taking antiviral medication) whose viral load has been suppressed

---

Take a look at the table on spousal violence, extracted from a model dataset created by the Demographic and Health Surveys (DHS) Program of ICF International for training purposes (ICF, n.d.).

DOMESTIC VIOLENCE

Table DV.1 Experience of physical violence

Percentage of women age 15–49 who have ever experienced physical violence since age 15 and percentage who have experienced violence during the 12 months preceding the survey, by background characteristics, Model DHS 6 data

Background characteristic	Percentage who have ever experienced physical violence since age 15[1]	Percentage who have experienced physical violence in the past 12 months			Number of women
		Often	Sometimes	Often or sometimes[2]	
Age					
15–19	47.6	5.7	24.1	29.9	408
20–24	56.2	4.9	23.1	28.1	388
25–29	56.1	7.2	26.1	33.7	450
30–39	59.3	8.2	22.2	30.5	790
40–49	57.8	2.8	15.2	18.4	560
Religion					
Religion 1	53.5	3.7	24.1	27.7	583
Religion 2	57.0	6.7	21.3	28.1	1,999
Religion 3	*	*	*	*	5
No religion	*	*	*	*	3
Ethnic group					
Ethnic group 1	52.3	3.5	24.3	27.8	603
Ethnic group 2	52.1	6.1	20.7	27.1	813
Ethnic group 3	50.8	6.2	26.1	32.3	119
Ethnic group 4	63.5	6.8	21.1	28.1	886
Ethnic group 5	56.9	11.5	20.6	32.1	153
Other	*	*	*	*	12
Missing	*	*	*	*	10
Residence					
Urban	58.0	6.0	21.9	28.1	947
Rural	55.0	6.0	21.7	27.9	1,649
Region					
Region 1	61.4	6.7	21.9	28.6	949
Region 2	53.2	4.7	21.1	26.2	521
Region 3	58.4	6.8	21.1	28.2	602
Region 4	46.8	5.2	23.2	28.4	524
Marital status					
Never married	47.5	5.3	17.4	22.7	511
Married or living together	57.5	5.9	23.5	29.6	1,900
Divorced/separated/widowed	66.1	9.4	16.4	25.8	185

Number of living children					
0	43.8	6.2	19.5	25.6	475
1–2	59.9	6.8	26.2	33.0	840
3–4	57.2	6.0	22.0	28.4	710
5+	59.5	4.7	17.1	22.1	571
Employment					
Employed for cash	62.6	6.6	23.7	30.3	861
Employed not for cash	55.1	6.1	21.5	27.7	1,130
Not employed	48.8	5.1	19.7	25.2	600
Missing	*	*	*	*	6
Education					
No education	55.9	5.5	22.4	28.1	1,607
Primary	60.2	9.4	20.7	30.6	317
Secondary or higher	54.8	5.5	21.0	26.5	673
Wealth quintile					
Lowest	49.7	5.4	19.4	24.8	534
Second	55.2	5.4	24.4	29.8	524
Middle	58.0	6.4	21.9	28.8	490
Fourth	58.5	7.0	23.3	30.3	509
Highest	59.4	5.9	20.3	26.6	539
Total 15–49	56.1	6.0	21.8	28.0	2,596

Note: An asterisk indicates that a figure is based on fewer than 25 unweighted cases and has been suppressed.

[1] Includes violence in the past 12 months. For women who were married before age 15 and who reported physical violence by a spouse, the violence could have occurred before age 15.

[2] Includes women for whom frequency in the past 12 months is not known.

The title of the table implies that the figures you see are percentages. Let's select the column with the title "Percentage who have ever experienced physical violence since age 15". Scroll down the column. The first figure you encounter is 47.6. From this figure, if you scroll leftwards, you will notice that the row header is "15–19". Thus, 47.6% (or nearly half) of women aged 15–19 years have experienced physical domestic violence. The next four figures below 47.6 show the percentages for other age groups. These percentages are high and potentially newsworthy.

Now, scroll down the table until you find the row header 'Region 1'. From this row header if you scroll right, the first figure you encounter is 61.4. In other words, 61.4% of the women in Region 1 have experienced physical domestic violence since the age of 15. The next three figures below 61.4 show the percentages for other regions. By comparing the percentages, you will

---

Learning activity 9.9: Spot the story

Look at the domestic violence table again. Once again, look at the column with the header "Percentage who have ever experienced physical violence since age 15". Scroll down and take note of the highest figure in that column.

1. What is the highest figure in the column with the header "Percentage who have ever experienced physical violence since age 15"?

2. What does that figure represent? What do you conclude?

Please write down your response before reading on.

Answers

1. The highest figure is 66.1.

2. It represents the percentage of divorced, separated or widowed women who have ever experienced physical violence since age 15. Divorced, separated or widowed women have experienced more physical violence than those who are currently married, living together or have never been married.

Learning activity 9.10: Women's employment status

Take a look at the domestic violence table again. Identify the row headers for employment status. Find the physical violence figures for the different employment categories.

With regard to employment status, which category of women has experienced more physical violence?

a) Employed for cash

b) Employed not for cash

c) Not employed

Please write down your response before reading on.

Answer

a) Employed for cash

Do you see a story here? If employed women experience more domestic violence than unemployed women, does that meet the criteria for what makes news? Is it surprising? Is this worth further investigating? Yes it is.

notice that Region 1 leads in physical domestic violence. Region 4 has the lowest percentage of women who have ever experienced physical domestic violence.

How to move from numbers to an interesting and informative story

So far, all you've got from the table is statistics that point you to a story. Remember, a journalist is a storyteller and not a statistician, therefore just reporting the numbers won't be sufficient. As a journalist you should focus more on people than on numbers. At the end of the day a story derived from statistics has to meet the criteria for a good science story: it should be easily understandable, relatable and accurate. In other words, having obtained the statistics, you need to carry out a series of interviews and observations to build your story.

Learning activity 9.11: Adding human voices

In regard to the domestic violence story identified above, write down a list of people you would want to interview in order to get a comprehensive story.

Please write down your response before reading on.

Feedback

- A past or present victim of domestic violence
- The police
- Employers
- Human rights campaigners
- A marriage counsellor
- A lawyer
- Children affected by parents' spousal violence
- A researcher involved in producing the statistics

Summary

- Technical reports are a rich source of stories for the news media.
- Reading the executive summary of a report is a good way to get the main points of the report at a glance.
- Numbers and statistics help us to understand the world around us better.
- Statistical tables can give you a headline or valuable information to substantiate a story.
- Look carefully when comparing the figures in the different cells, columns and rows of a table: a journalist can spot a great story in statistical tables.
- Getting the facts and statistics from a technical report is only a gateway to a story. You need to build the story by adding the experiences, views and expertise of human beings.

References

CABI (2017) *Invasive Species: The Livelihoods Threat: Progress report on pre-implementation activities for the CABI invasive species programme in Africa and Asia, 2014–2016.* Available at: https://www.cabi.org/wp-content/uploads/projectsdb/documents/62665/CABI % 20Invasives % 202014-2016 % 20Report.pdf (accessed 17 February 2022).

FAO (2018a) *The State of the World's Forests: Forest pathways to sustainable development.* Available at: www.fao.org/3/I9535EN/i9535en.pdf (accessed 22 July 2022).

FAO (2018b) Integrated management of the fall armyworm on maize: A guide for farmer field schools in Africa. Available at http://www.fao.org/3/I9535EN/i9535en.pdf (Accessed 17 February 2022).

ICF (n.d.) Model datasets. Funded by USAID. Available at: https://www.dhsprogram.com/data/Download-Model-Datasets.cfm (accessed 21 March 2022).

Ombuor, R. (2018) Vasectomies slowly gaining acceptance in Kenya. VOANews.com 28 May. Available at: https://www.voanews.com/a/vasectomies-slowly-gaining-acceptance-in-kenya/4413227.html (accessed 29 April 2021).

Scott, K. (2019) Can the Middle East solve its water problem? *CNN.com,* 22 March. Available at: https://edition.cnn.com/2018/07/11/middleeast/middle-east-water/index.html (accessed 29 April 2021).

UNAIDS (2017) *Data 2017.* Available at: https://www.unaids.org/sites/default/files/media_asset/20170720_Data_book_2017_en.pdf (accessed 17 February 2022).

UNESCO (2010) *UNESCO Science Report 2010: The current status of science around the world.* Available at: https://unesdoc.unesco.org/ark:/48223/pf0000189958_eng (accessed 29 April 2021).

VOA News (2018) 42 percent of Americans say teen sex is "morally acceptable". *VOA News* 11 July. Available at: https://learningenglish.voanews.com/a/percent-of-americans-say-teen-sex-morally-acceptable-/4476713.html (accessed 29 April 2021).

World Bank (2018) *Beyond Scarcity: Water Security in the Middle East and North Africa.* MENA Development report. Available at: https://openknowledge.worldbank.org/handle/10986/27659 (accessed 29 April 2021).

Advanced Technical Skills for Science Reporting

10

Most journalists get their science stories from press releases and then add a few comments from human sources. However, that may not be enough to give you an award-winning story. Additional skills can help journalists to report exclusive science stories. Key among these are the skills of independently interpreting research findings and getting a good story from original data.

Interpreting research results

Accurate science reporting begins with an accurate interpretation of the research findings. The key question is: "What conclusion can I make from these results?" Correct interpretation of the findings leads to the right conclusions and therefore an accurate story. Inaccurate interpretation of research findings inevitably leads to an inaccurate story.

Press releases and errors in the news

Press releases are the most common source of science stories for journalists. Most often, journalists base their story on the facts presented in a press release, without reading the research paper to which it relates. However, research studies show that press releases can mislead journalists and lead to errors in the news media. One such study is highlighted in the article "Bad science reporting blamed on exaggerations in university press releases" (see https://www.independent.co.uk/news/science/bad-science-reporting-blamed-on-exaggerations-in-university-press-releases-9913336.html) (Connor, 2014). The study, published in the *British Medical Journal* in 2014, examined 462 press releases issued by 20 British universities. In summary, the findings were that:

- 36% of the press releases wrongly implied that the results were applicable to human beings, yet the studies were actually only done with animals
- 40% of the press releases added health advice that was not in the research paper concerned

- 33% of the press releases made false claims, saying that one thing caused another, when there was no real evidence for this and these claims were not in the original research paper.

Learning activity 10.1: Exaggerations in press releases

The study referred to above clearly shows that press releases often include exaggerations and even false claims about research findings. Write down the main problems that exaggerations in press releases can cause, and what journalists can do to prevent it.

Please write down your response before reading on.

Feedback

Exaggerations in press releases lead to inaccurate stories in the media, which can mislead the public and policymakers into making wrong – and perhaps dangerous – decisions. Therefore, journalists should not rely entirely on press releases, which might bias or mislead them. Beyond the press release, science journalists should read the original research paper to which it relates, and make their own interpretations of the findings before carrying out interviews.

Understanding research results – key terms and concepts

There are several terms and concepts that you need to know in order to interpret research findings, such as **mean**, **sampling**, **p-value** and **sampling error**.

Learning activity 10.2: Calculating the mean

Look at the table below, which gives the body weights of six members of a hypothetical sports team called "Shooting Stars".

Body weights of Shooting Stars team members	
Name	Weight (kg)
Isaac	67
Chima	60
Azizi	65
Bekele	63
Peter	70
Nyongo	65
Average weight	

Calculate their average body weight (the sum of all of the weights divided by six). All you need to do this is a simple calculator.

Please write down your response before reading on.

Feedback

The average, which scientists refer to as the **mean**, is 65. Write this down, because we're going to refer to it later in the chapter: we call it the **whole group mean** or the **population mean**.

Learning activity 10.3: Sampling

Using a random process, pick three out of the six names in the table: this is your **sample**. You can use an online randomizer to do this, such as Random Lists (see https://www.randomlists.com/list-randomizer?items=Isaac % 0AChima % 0AAzizi % 0ABekele % 0APeter % 0ANyongo % 0A&qty=3&dup=false). Simply enter the names into the randomizer (or copy and paste them if you are using the e-book version of this book) and specify your sample size (the number of names you want the randomizer to select). Alternatively, you can write each name on a separate piece of paper, roll them up, mix them in a bowl, close your eyes and pick out three at random.

Find the total of their body weights and divide it by three. This is your **sample mean**. Write down the figure.

Now, compare your whole group mean and the sample mean. What do you see? Is the sample mean the same as the whole group mean?

Please write down your response before reading on.

Feedback

The mean of a sample is usually not exactly the same as that of the entire population. This is because the population is not homogeneous. Due to individual differences, a sample is never a perfect representation of the entire population.

Although a sample is not a perfect representation of a population, in most cases scientific researchers cannot avoid sampling. Let's say a researcher wants to study social media use among students at a university. It is not feasible to ask every university student about their social media use, so the researcher will administer the questionnaire to a sample of students. This leads us to another concept: **sampling error**.

Sampling error

As you have seen in Learning activity 10.3, the **sample mean** will not be exactly the same as the **population mean**. This difference is called **sampling error**. This means that, when reporting on study results, journalists need to pay attention to the sampling error. Generally, the bigger the sample, the more the results will be representative of the general population. Likewise, the more random the sampling process is, the more representative the results will be.

Therefore, before generalizing the results, a journalist should ask some questions. Is the sample size big enough to represent the population? Is the sampling method credible enough to represent the population? If, for example, you administer a questionnaire only to your classmates because this is convenient for you, and you find that 20% of them watch a certain TV programme, you should not generalize the results to say that 20% of students at your university watch that TV programme. At best, those results can only be representative of your class, not the university. Similarly, if you went out to interview 20 out of 20,000 students at a university, you wouldn't be able to generalize your results to represent the whole campus, because the sample size is too small.

Learning activity 10.4: P-value and statistical significance

Let's carry out another exercise with the Shooting Stars team.

Body weights of Shooting Stars team members

Name	Weight (kg)
Isaac	67
Chima	60
Azizi	65
Bekele	63
Peter	70
Nyongo	65
Average weight	

Randomly divide the Shooting Stars team into two groups of three. As in Learning activity 10.3 you can use an online randomizer or a manual method. Calculate the mean weight of each group and write it down. Create a table and fill in the mean body weights of the two groups and the whole team, as shown below.

Category	Mean body weight
Shooting Stars (whole team)	65
Group 1	XX
Group 2	XX

What do you see? Are the mean weights the same?

Please write down your response before reading on.

Feedback

The mean for Group 1 is almost certainly different from that for Group 2. Due to the inherent differences between individuals, two subsets of a population are normally not identical. This means that in scientific research some of the differences observed between two subgroups of a population might be purely the result of chance. This leads us to the concept of **p-value**.

What is p-value?

Because of the individual differences explained above, researchers always assume there is a small probability that some of the results they obtain are purely the result of chance. That probability is called the **p-value**. The p-value is measured on a scale of 0 to 1. Usually it is calculated automatically by statistical programmes. The p-value depends on the sample size and the variability of the population.

Let's say a researcher wanted to establish which of two feed formulations results in faster growth for poultry. They would divide the birds into two groups, assign each group a different feed formulation and compute the mean weekly weight gain for each group to see which one leads to faster growth. If the p-value is high, the researcher concludes that the difference observed between the two groups is most likely the result of chance rather then the different feeds given to them. If the p-value is low, then the differences observed are most likely real.

The question is, how high is too high? Most researchers set the cut-off point at either 0.01 or 0.05. This cut-off point is called the **level of significance**.

To better understand the level of significance, let's convert the decimals to percentages using a calculator or an online converter. Better still, here's a maths trick: simply shift the decimal point by two positions to the right: for example 0.005 becomes 0.5% and 0.15 becomes 15%.

Learning activity 10.5: Decimals and percentages

Convert the following decimals to percentages.

0.001

0.02

0.08

0.17

0.3

Please write down your response before reading on.

Feedback

Decimal	Percentage
0.001	0.1%
0.02	2%
0.08	8%
0.17	17%
0.3	30%

Implications of the level of significance

A p-value of 0.05 means there is a 5% chance, or a 1 in 20 chance that the results you are observing are purely the result of chance or coincidence. Similarly 0.001 is the same as 0.1%, so a p-value of 0.001 represents a 1 in 1,000 chance that the differences you see between the two groups are coincidental.

If the p-value is higher than the level of significance, the results are considered to be **not statistically significant**. This means there is too high a probability that the differences observed between the two groups are purely the result of chance. If the p-value is lower than the level of significance, the results are considered to be **statistically significant**.

Note that this is different from the layperson's understanding of the word "significant". In everyday language significant means important, while in statistics the term significant is related to the probability that the difference seen between two groups is real.

Therefore, when reporting on scientific research, pay attention to whether the results are statistically significant. In quantitative studies, the researchers will normally state the p-value in their research paper, and whether the findings are statistically significant.

Moving from the results section of a research paper to headlines

Having discussed the different terms and concepts used in presenting research results, let's now focus on translating those results into a story.

Correlation is not causation

In scientific research the terms "correlation" and "association" are used to refer to two things that are connected to each other. However, when two things always happen concurrently, it doesn't mean one causes the other: for example, during sunny and hot seasons, more people are likely to wear sunglasses and to eat ice cream. It then appears that the more people wear sunglasses the more they eat ice cream.

Learning activity 10.6: Sales of ice cream and sunglasses

Why would sales of ice cream and sunglasses increase concurrently during certain periods? Select the most probable answer:

a)   Sunglasses make people want to eat more ice cream.

b)   Ice cream makes people want to wear glasses.

c)   The demand for ice cream and sunglasses are both influenced by warmer, sunny weather.

Please write down your response before reading on.

Feedback

The correct answer is option (c), the demand for ice cream and sunglasses are both influenced by warmer, sunny weather.

There is a correlation between wearing sunglasses and eating ice cream. That doesn't mean that wearing sunglasses causes people to crave ice cream. Neither eating ice cream nor wearing sunglasses causes the other, but they are both related to something else: namely hot, sunny weather. That is what we mean when we say that *correlation is not causation.*

Does eating eggs prevent heart disease?

The study reported on in the article "Associations of egg consumption with cardiovascular disease in a cohort study of 0.5 million Chinese adults" (see https://pubmed.ncbi.nlm.nih.gov/29785957/) (Qin *et al.*, 2018) is another example of a case where correlation is not causation. The researchers followed up about half a million adults, categorized them according to how often they ate eggs and then monitored them for several years. They observed that those who ate eggs daily were at a lower risk of heart disease. Their conclusion was as follows: "Among Chinese adults, a moderate level of egg consumption (up to <1 egg/day) was significantly associated with lower risk of CVD [cardiovascular disease], largely independent of other risk factors" (Qin *et al.*, 2018).

Learning activity 10.7: Eggs and heart disease

Below are some of the news headlines published by various media outlets as a result of the above study.

Headline 1: "Egg a day tied to lower risk of heart disease" (see https://www.reuters.com/article/us-health-heart-eggs/egg-a-day-tied-to-lower-risk-of-heart-disease-idUSKC-N1IM2G9) (Rapaport, 2018).

Headline 2: "An egg a day might reduce your risk of heart disease, study says" (see https://edition.cnn.com/2018/05/21/health/eggs-heart-disease-study/index.html) (Scutti, 2018).

Headline 3: "Eating an egg a day may keep heart disease away, a new study says" (see https://time.com/5285672/eggs-lower-heart-disease-stroke/) (Park, 2018).

Headline 4: "Eating an egg a day can reduce the risk of heart disease and stroke" (see https://inews.co.uk/news/health/eating-daily-egg-reduce-risk-heart-disease-stroke-156791) (Gallagher, 2018).

Headline 5: "Eating eggs 'significantly' cuts heart disease risk" (see https://www.foodnavigator.com/Article/2018/05/22/Eating-eggs-significantly-cuts-heart-disease-risk) (Askew, 2018).

Continued

Learning activity 10.7. Continued.

Based on the description of the study and the conclusion, discussed above, which of these five headlines are accurate reflections of the study findings? Which ones imply a causation where there is only a correlation?

Please write down your response before reading on.

Feedback

Headline	Comment
"Egg a day tied to lower risk of heart disease"	This is an accurate reflection of the study findings.
"An egg a day might reduce your risk of heart disease, study says"	The writer tried to play it safe using the word "might". Nevertheless, some readers will perceive a causation.
"Eating an egg a day may keep heart disease away, a new study says"	The writer tried to play it safe using the word "may". Nevertheless, some readers will perceive a causation.
"Eating an egg a day can reduce the risk of heart disease and stroke"	"Can" is more emphatic than "may" and "might", so this headline is more misleading than the previous two.
"Eating eggs 'significantly' cuts heart disease risk"	This is definitely inaccurate. It implies that eating eggs definitely lowers the risk of cardiovascular disease.

The researchers knew they did not have enough evidence to say eating eggs reduces the risk of getting cardiovascular disease. That is why they use the phrase "associated with" and not "reduces" or "causes". Furthermore, the paper has a limitations section where the researchers say they cannot rule out the possibility of other factors having contributed to the reduced risk of cardiovascular disease.

Experimental vs non-experimental studies

The study above, on eggs and cardiovascular disease, is a **non-experimental study**. Non-experimental studies do not prove that one thing causes another. There are several types of non-experimental research, such as observational studies, correlational studies or cross-sectional studies (refer to Chapter 1). The bottom line is that they are not experiments and therefore they cannot prove that one thing causes the other. What the eggs study did was to confirm that there is a correlation between eating eggs and lowered risk of cardiovascular disease. A different kind of study is needed to establish whether eating eggs can indeed reduce the risk of cardiovascular disease.

Another example is an observational study in northern Uganda, which found that households that owned goats suffered fewer cases of malaria (Maziarz *et al.*, 2018). That doesn't mean goats protected them from malaria. Instead, there could be another factor related to both owning a goat and prevention of malaria. In rural Uganda owning livestock is an indicator of socioeconomic status and, therefore, households that have livestock are more likely to afford mosquito nets.

The absence of evidence is not evidence of absence

Some things are obvious to the human senses: we can see, feel or hear them. If a teacher is in the classroom you will see and hear them. Some things are not as obvious as that. Take, for example, a microscope-based malaria test: the lab technician will not report that the patient has no malaria parasites. Instead, the technician will report that they have not *seen* malaria parasites in the sample. This is because there is a slim chance that the malaria parasites exist but that the technician did not detect them.

Take a look at the quote below, extracted from the article "Sorry guys: the most comprehensive radio search ever just found no sign of alien life" (see https://www.sciencealert.com/the-most-comprehensive-and-sensitive-radio-search-ever-has-found-no-sign-of-alien-life) (Cassella, 2019)

> We scoured thousands of hours of observations of nearby stars, across billions of frequency channels. We found no evidence of artificial signals from beyond Earth, but this doesn't mean there isn't intelligent life out there: we may just not have looked in the right place yet, or peered deep enough to detect faint signals.

Researchers use phrases like "no evidence", "no significant difference" and "we did not find …" deliberately. Where you see such phrases, don't assume that a researcher's failure to detect something means that thing doesn't exist.

Three other issues to look out for

Animal experiments

Often, new medical treatments – for example, vaccines and drugs – are tested on small mammals, such as mice, to ensure their safety and effectiveness before they're tested on human beings. However, the results can be quite misleading because mice are simply not human beings. If scientists find that garlic kills slugs it doesn't mean that garlic kills human beings. If a new experimental drug is effective and safe in treating mice it doesn't mean the same result will be obtained in human beings. Nevertheless, animal studies help scientists to know whether they need to carry out a similar experiment with human beings. They should therefore be reported in that context.

Computer models

Computer modelling is used in various fields, including weather forecasts, earthquake prediction, mapping disease patterns and clinical trials. For instance, increasingly, scientists are using computer models to predict the probable effect of an experimental drug in human beings. But human beings are not computer programmes, so some of the findings might deviate from reality. Therefore, findings based on computer modelling should be reported in that context.

Initial findings

In Botswana, a team of researchers noted that 4 out of 426 women had produced babies with neural tube defects after taking the new antiretroviral drug Dolutegravir (DTG) for some time. The WHO then advised that DTG should not be given to women of child-bearing age until further research was done on its safety. However, further studies could not find any link between DTG and neural tube defects. As a result, the WHO adjusted its position and endorsed the use of DTG among women of child-bearing age.

Quite often, researchers report preliminary or initial findings that require additional research to validate them. In some cases additional research can disapprove the findings of the earlier research. Avoid publicizing such initial findings but, if you must report them, then specify that those results still need to be validated.

Data mining and visualization

Data journalism takes a journalist's role to a new level. Instead of just reporting on a scientist's research findings, you can access data online, analyse the data and write an interesting story.

What is data?

You will come across different definitions of "research data", using different words, but most definitions define data as a set of facts that have been collected and stored in a manner that enables them to be accessed and analysed by researchers.

Data by itself, for example a recording of all people present in a country on census night, and their age, may not mean much to a non-trained person looking at it cursorily. But by identifying and describing trends and patterns in the data, you can generate useful information for your audience. For example, from the above-mentioned census data you can compute the average age per county, and therefore figure out where people live longest. You could also compare these figures with previous censuses to spot a trend.

Similarly, a recording of the number of people who tested HIV positive in a given country does not say much on its own. However, an important story can emerge when you compare different years and regions, gender or age groups, in order to see whether the number is declining or increasing, and which area is most affected. It also makes sense to compare the number with the targeted or expected figure, to determine whether the country is on course to improve the situation.

How to find free data

Today, more data is available online than ever before; the trick is to know where and how to find reliable data.

Learning activity 10.8: Find datasets online
This activity requires you to have access to the internet.

1. Select a topic that you want to research: for example, "forest cover in Africa".
2. Go to Google dataset search (see https://datasetsearch.research.google.com/).
3. In the search box, enter your search terms (e.g. "forest cover in Africa").
4. In your search results, scroll down and select the dataset you want (e.g. "global forest canopy density – Africa").
5. Select and download the file (save the file because you will need it for the next learning activity.)

Alternatively, instead of a Google dataset search, you could go straight to known sources of free datasets, such as:

- Global Health Observatory Data Repository (see https://apps.who.int/gho/data/node.home)
- World Bank Open Data (see https://data.worldbank.org/)
- Academic Torrents (see http://academictorrents.com/)
- FAOSTAT – Data (see http://www.fao.org/faostat/en/#data).

Is the data clean?

Data cleaning involves removing incorrect, incomplete, duplicated or wrongly formatted data. It goes without saying that uncleaned data will give you faulty results – and therefore misleading interpretations. So, before proceeding with analysis, you need to "clean" the data. If the data was collected and stored by someone else, you need to find out whether it was cleaned. There are many online data cleaning tools: for example OpenRefine (see https://openrefine.org/).

Data visualization

Data journalists use visuals to produce informative and captivating stories. They represent data through charts and graphs. The aim is to communicate science in a way that is appealing and interesting for the viewer or reader.

Learning activity 10.9: Visualizing your data

This activity requires you to have access to the internet.

1. Open Google Sheets (see https://www.google.com/sheets/about/).
2. Select "Blank" to start a new spreadsheet.
3. Import the file you downloaded in Activity 10.8 (File, Import).
4. Select the data you want to analyse.
5. On the menu bar at the top, click "Insert", then select "Chart".
6. Double-click on the chart to reveal a chart editor. Under "Chart Type", click on the downward arrow to select the chart type you want (pie chart, bar chart, column, geo, etc.).
7. Save your chart as a PDF or PNG file.

Other tools you can use for visualizing data include:

- Flourish (URL: https://flourish.studio/)
- Datawrapper (URL: https://www.datawrapper.de)
- Microsoft Excel.

Some sources of data have in-built provisions for analysis and visualization. In that case you might not need to go to Google Sheets for analysis and visualization. Please note that technology changes at a fast rate so by the time you read this book newer tools or methods might be available. As a journalist, it is important to keep abreast of new technologies.

Benefits of data visualization

Data visualization has the following advantages:

- visualization can help your audience to easily comprehend the information
- it gives your audience a quick snapshot of the story, without necessarily going through your entire narrative
- it can aid memorability – people remember what they have seen better than what they have only heard or read
- well-done visualization will make your story look more authoritative in the eyes of your audience.

Summary

- Stories about research findings influence policy and personal decisions, therefore it is important to base stories on a correct interpretation of research findings.
- When reading research papers it is important to pay attention to sampling, p-values and level of significance.
- When two variables increase or reduce concurrently, it doesn't mean one causes the other. Correlation is not causation.
- The results of animal studies should be reported in context so as not to mislead people, because they cannot be extrapolated to human beings.
- If you must report about initial findings that still need validation, it is important to let your audience know the research results are not conclusive.
- Apart from reporting on published research findings, journalists can access data online, and analyse and visualize it, to produce an exclusive story.
- Ensure you verify and cross-check facts you source from online databases.

References

Askew, K. (2018) Egg consumption "significantly" cuts heart disease risk, study suggests. Foodnavigator.com 22 May. Available at: https://www.foodnavigator.com/Article/2018/05/22/Eating-eggs-significantly-cuts-heart-disease-risk (accessed 30 April 2021).

Cassella, C. (2019) Sorry guys: The most comprehensive radio search ever just found no sign of alien life. Sciencealert.com 22 June. Available at: https://www.sciencealert.com/the-most-comprehensive-and-sensitive-radio-search-ever-has-found-no-sign-of-alien-life (accessed 30 April 2021).

Connor, S. (2014) Bad science reporting blamed on exaggerations in university press releases. Independent.co.uk 10 December. Available at: https://www.independent.co.uk/news/science/bad-science-reporting-blamed-exaggerations-university-press-releases-9913336.html (accessed 30 April 2021).

Gallagher, P. (2018) Eating an egg a day can reduce the risk of heart disease and stroke. Inews.co.uk 22 May. Available at: https://inews.co.uk/news/health/eating-daily-egg-reduce-risk-heart-disease-stroke-156791 (accessed 30 April 2021).

Maziarz, M., Nabalende, H., Otim, I., *et al.* (2018) A cross-sectional study of asymptomatic Plasmodium falciparum infection burden and risk factors in general population children in 12 villages in northern Uganda. *Malaria Journal* 17, 240. Available at: https://malariajournal.biomedcentral.com/articles/10.1186/s12936-018-2379-1 (accessed 17 February 2022).

Park, A. (2018) Eating an egg a day may keep heart disease away, a new study says. Time.com 21 May. Available at: https://perma.cc/8HQW-GCWT (accessed 30 April 2021).

Qin, C., Lv, J., Guo, L., Bian, Z., Si, J., Yang, L., Chen, Y., *et al.* (2018) Associations of egg consumption with cardiovascular disease in a cohort study of 0.5 million Chinese adults. *Heart* 104, 1756–1763.

Rapaport, L. (2018) Egg a day tied to lower risk of heart disease. Reuters.com 22 May. Available at: https://perma.cc/MFL9-9EK4 (accessed 30 April 2021).

Scutti, S. (2018) An egg a day might reduce your risk of heart disease, study says. Edition.CNN.com 21 May. Available at: https://edition.cnn.com/2018/05/21/health/eggs-heart-disease-study/index.html (accessed 30 April 2021).

Reporting Science Responsibly

In this chapter

- Responsible journalism
- Ethics of reporting scientific evidence
- Ethical principles of research and journalism
- Applying ethical principles
- Covering controversies in science
- Potential risks of controversies in science
- Reporting on controversial issues

Responsible journalism

Responsible journalism requires an understanding of the essence of journalism and its ethical principles, and respectable conduct in gathering and packaging the story, in order to produce trustworthy media content. Any deviation from responsible journalism amounts to a betrayal not only of society, but also of the media. When individuals stop trusting the media, they begin to avoid the news, which works against the media. Therefore, reporting science responsibly is in the interest of both society and the media fraternity.

Learning activity 11.1: The purpose of journalism

Why do journalists report stories? Look at the following statements regarding the purpose of journalism and consider what they have in common.

Journalism is giving people the best possible version of the truth in order for people to make up their minds.

Ulrik Haagerup, founder of the Constructive Institute ("What is the purpose of journalism today?", https://agency.reuters.com/en/insights/videos/what-is-the-purpose-of-journalism-today.html)

The purpose of journalism is thus to provide citizens with the information they need to make the best possible decisions about their lives, their communities, their societies, and their governments.

American Press Institute (see https://www.americanpressinstitute.org/journalism-essentials/what-is-journalism/purpose-journalism/)

Continued

©2022 CAB International. Science Communication Skills for Journalists: A Resource Book for Universities in Africa (Ed. Charles Wendo)
DOI:10.1079/9781789249675.0011

Learning activity 11.1. Continued.

It is nearly impossible to say journalism has one sole purpose. Ensuring society becomes well informed and providing it with reliable information is incredibly important. It is also a journalist's duty to deliver accurate news, true to the source without being biased or taking sides.

Open School of Journalism

Write down the key points that at least two of these definitions have in common.

Please write down your response before reading on.

Feedback
- Society (also referred to as people or citizens)
- Information (one statement refers to it as truth)
- Credible, reliable
- Making decisions (also referred to as making up their minds)

Journalists provide people with the credible information they need for making informed decisions about their lives and societies. Closely related to this is the watchdog role, whereby journalists monitor, and expose, hidden information about the conduct of powerful people, especially government officials.

In a nutshell, the decisions we make at the personal, family, community or national level are often informed by the information we get from the media and other sources. Unlike many other sources of information online, a journalist's story goes through a process of verification that makes it credible – this is the basis on which audiences trust the news media.

Ethics of reporting scientific evidence

Underpinning arguments for responsible journalism is the fact that the media are accountable to society. But before we proceed, let's try to reach a common understanding of ethics.

Ethics is about deciding – and then doing – what's right in your course of work. The ethics of reporting scientific evidence is about the acceptable standards of conduct that a journalist must follow in communicating research.

Learning activity 11.2: What does ethics mean to you?

Write down at least one word or phrase which sums up what ethics means to you.

Please write down your response before reading on.

Feedback
- Values
- Principles
- Good conduct
- Good practice
- Doing right

Ethical principles of research and journalism

Because your role involves reporting science, you need to be aware of the ethics of both research and journalism. The various codes of research ethics have the following in common:

- a commitment to scientific truth
- being accountable to other researchers and society
- doing research in a way that maximizes benefit and minimizes harm
- an obligation to disseminate research findings
- confidentiality of participant information
- respecting the privacy and confidentiality of research subjects
- scientific integrity
- fairness and justice
- carrying out research that does not promote prejudice about, and stigma towards, specific social groups and individuals.

Like scientific research, journalism is guided by a set of ethical principles as illustrated in Fig. 11.1. Below are the ethical principles put forward by the Society of Professional Journalists (SPJ) (2014) (see https://www.spj.org/ethicscode.asp).

- **Seek the truth and report it** – Journalists should be courageous in seeking information. They should interpret the information correctly and report it accurately, fairly and honestly.
- **Minimize harm** – Journalists should avoid causing harm to the community and their sources of information.
- **Act independently** – Journalists should serve the interests of the public and avoid conflicts of interest.
- **Be accountable and transparent to society** – This includes acknowledging and correcting mistakes, and responding promptly to complaints from the public.

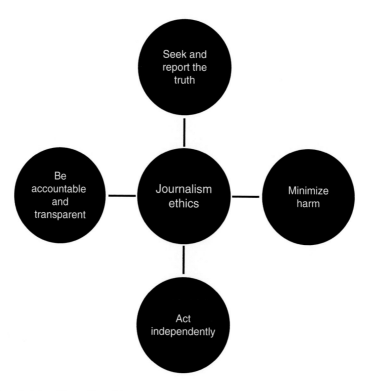

Fig. 11.1. Key principles of journalism ethics.

> **Learning activity 11.3: The ethics of research and the ethics of journalism –**
> **common ground**
> Compare the ethical principles for scientific researchers with those for journalists (as listed
> above) and identify the similarities. Write down the principles that apply to both science
> *and* journalism.
>
> *Please write down your response before reading on.*
>
> **Feedback**
> - Truth
> - Minimizing harm
> - Fairness and justice
> - Being accountable to society

Applying ethical principles

Here are some ways that you can apply the above ethical principles to your work of reporting science:

Respect embargoes

A number of scientific journals release research results under **embargo**. This could be in the form of a press release, an abstract of the study or the full study. A journalist should not publish a story on the study before the date the embargo ends.

Avoid plagiarism

Do not copy and paste text from press releases and research papers into your story. As a journalist, you have to understand the study findings and repackage them. Attribute information to sources. In science, every fact has to be supported by evidence attributed to the source or owner.

Selection of interviewees

Whether you are reporting about research findings, policy pronouncements or any new development that has scientific implications, you need to interview a number of people to get a complete story. It is important to know who to interview for what. For example, the inventor of a piece of technology is not the right person to ask when trying to find out if the target communities like it, neither would you rely on a policymaker or a random community member if you have questions regarding the scientific credibility of research. Some interviewees you might seek out include the following.

- **The experts** – This might include the author of a study you are reporting on, as well as an independent expert on the same subject. The independent expert may express agreement or disagreement with the key player. Later in this chapter we will discuss how to handle scientific controversy ethically.
- **An affected person** – You may want to ask them how the issue is affecting them or how it is likely to affect them, and whether they find it agreeable or positive. Remember, something might be scientifically sound and safe but people may reject it, for various reasons – perhaps for cultural or religious reasons or for reasons of convenience.

- **A policymaker** – You may want to interview them to find out about the policy and programme implications. Be mindful that a scientifically sound pronouncement can be rejected by policymakers on the grounds that it might be expensive, unpopular or difficult to implement, or even for selfish reasons.
- **An implementer** – This could include officials of a government department or a representative of a non-governmental organization working on community development.

Conflicts of interest

Your readers, listeners or viewers must know whether there are any conflicting interests in the execution of a study. Find out how the study has been funded and who is behind it. Checking the affiliation of the researchers could also help in determining whether there is any conflict of interest. Some studies have a specific section for authors to declare conflicts of interest; however, you may have to delve deep to find out any potential conflict of interest.

As a reporter, you should also state if you were funded to report on research evidence: for example, a paid trip to a field trial site or being given airtime.

Learning activity 11.4: Retracted paper

Read the summary below.

> In 2018 a team of American researchers published a paper (see https://www.nature.com/articles/s41586-018-0651-8) in *Nature* indicating that oceans were warming faster than previously thought (Resplandy *et al.*, 2018). Their paper received wide media coverage. Following criticism by other scientists, the researchers notified *Nature* about potential problems with their method of analysis. After further scrutiny, *Nature* retracted the paper on the basis that the researchers had overstated the level of certainty about their findings (see https://www.nature.com/articles/s41586-019-1585-5) (Resplandy *et al.*, 2019).

If you were the journalist who reported on this study the first time round, what would be your best course of action?

a) Avoid reporting on the retraction because it would embarrass you and your media house to publish another story stating something different from the first one.

b) Blacklist the scientist(s) and never report on any of their research in future.

c) Report on the retraction and make reference to your earlier story on the study.

d) Report on the retraction without making reference to your earlier story.

Please write down your response before reading on.

Answer and feedback

c) Report on the retraction and make reference to your earlier story on the study.

Remember, you are doing journalism in the public interest. The need to avoid embarrassment to the journalist or media outlet should never take precedence over the public interest. Research by Sarathchandra and McCright (2017) suggests that media coverage of a retraction can partly reverse the misinformation caused by an initial publication and its resultant media coverage (see https://journals.sagepub.com/doi/full/10.1177/2158244017709324).

Learning activity 11.5: Corrections

A journalist publishes a story correctly stating a scientist's research findings. A day later, the scientist emails to complain that the journalist made a mistake in explaining the social implications of the findings. From the options below, what is the journalist's best course of action?

a) Apologize to the scientist and take no further action.

b) Publish a correction, stating the correct facts and pointing out the error in the earlier story.

c) Publish an accurate version of the story without disclosing that you made a mistake in the earlier story.

Please write down your response before reading on.

Answer and feedback

b) Publish a correction, stating the correct facts and pointing out the error in the earlier story.

As explained earlier, it is an ethical responsibility of the journalist and media outlet to put the record straight. Fear of embarrassment should not prevent you from making a correction in the public interest. Acknowledging and correcting your mistake will actually enhance your reputation in the eyes of right-thinking people.

Learning activity 11.6: Limitations of a study

Read the excerpt below from a study on smallholder farmers' preference for weather index insurance in Kenya (Sibiko *et al.*, 2018, https://agricultureandfoodsecurity.biomedcentral.com/articles/10.1186/s40066-018-0200-6#Sec14).

> We caution that the results are specific to Kenya and that choice-experimental data may be subject to hypothetical bias. Hence, the exact estimates should not be generalized and over-interpreted. However, the findings still provide interesting insights into typical issues of WII design in a small-farm context. Given that smallholder farmers are particularly vulnerable to climate shocks, improving their access to crop insurance is of high policy relevance. More research is needed to further add to the knowledge base about suitable contractual designs in particular situations.

List how the above limitations would affect the way you report on the research findings.

Please write down your response before reading on.

Feedback

It is important to reveal that the study was carried out only in Kenya and that it is not clear whether the findings would be similar if the study was done in another country. It would be misleading to insinuate that the findings apply to all countries.

Covering controversies in science

In any discussion about what makes news, controversy features prominently. The media love controversy. But if a controversy is not reported accurately and responsibly, the story might cause more harm than good.

Learning activity 11.7: Defining controversy in science

What does controversy in science mean to you? Write down key words or phrases that define scientific controversy, for example disagreement.

Please write down your response before reading on.

Feedback
- Disagreement
- Public
- Scientific facts
- Prolonged

In a nutshell, scientific controversy is a prolonged disagreement among scientists over scientific knowledge.

Here are some of the common sources of controversy.

- **The very nature of science** – Activists opposed to a new technology (for example, genetically modified foods) will argue that the proponents cannot prove beyond doubt that it is safe. However, 100% certainty is not in the nature of science. Scientists usually provide for some degree of uncertainty in reporting their research findings.
- **Disagreement over the nature of research** – For example, some scientists and activists argue that it is wrong to carry out medical research using embryonic stem cells extracted from aborted foetuses.
- **Disagreements over the scientific evidence** – This arises when one group of scientists disputes the conclusions of another group's research. The dispute could be over the research methods used or the interpretation of the results.
- **Disagreement over the application of research findings** – For example, should people who have malaria in their blood but who are not sick be rounded up and treated against their will because they can be a source of infection for others?
- **Refusal to acknowledge the evidence due to intrinsic biases or vested interests** – This often happens as a result of religious beliefs and commercial interests. A good example is the tobacco industry's rejection of some of the evidence linking passive smoking to cancer.

Box 11.1. Common causes of scientific controversy.
- The nature of science – genuine uncertainty exists in science
- Disagreement over the nature of research
- Disagreement over the scientific evidence
- Disagreement over the application of research findings
- Refusal to acknowledge the evidence due to intrinsic biases or vested interests

Potential risks of controversies in science

- You need to watch out for controversial issues in science as they may distort communication with the public and undermine people's trust in science.

- Controversial issues can sometimes make it difficult for journalists to understand the issue being discussed, especially when they receive contradictory information on the same topic.
- Controversies can make it challenging to write a good story that has a clear message for your audience. You can easily end up writing a misleading story.

Reporting on controversial issues

Some of the most common controversial issues in science today include climate change, genetic engineering, vaccines, diets, stem cell research and space exploration. You need to strike a balance between the public's right to know and the risk of causing more confusion. Just because other journalists are covering or talking about the controversy does not necessarily mean you should follow suit.

Learning activity 11.8: To report or not to report?

A psychology researcher approaches you with the findings of his study – that people of a particular race are genetically more intelligent than others. How would you handle the information?

a) Immediately write the story and convince the editor that it is an important research finding.

b) Ignore the story because the research is controversial.

c) Report the story but add that the researcher has been criticized for his racial focus.

Please write down your response before reading on.

Feedback

This quote from Tim Radford, former science editor of the *Guardian* throws some light on this issue.

> Still, there are some 'science' stories that verge on lunacy. I was delighted some years ago to see dozens of US reporters simply walk out of a press conference at which a psychologist claimed to have established that some racial lineages were more intelligent than others. To report such a claim at all, even to dismiss it, would have provided bogus ammunition for some unwholesome political movements.
>
> (Radford, 2008)

Respect all legitimate voices

It's important to report fairly by giving airtime and space to the appropriate parties to air their views. Whereas the views of non-scientists are important in a story, ensure that debate on scientific facts is analysed by appropriate experts only. Your story should lead people to a point where they can make an informed decision. This implies that you should go beyond the contesting voices by doing more research, contextualizing the issues and determining when some perceptions or comments need to be ignored.

Go with the scientific consensus

Science is built on the collective judgement of specialists on a specific topic. This collective judgement is termed "scientific consensus". When there is consensus on an issue, it doesn't mean every scientist agrees – a few scientists will have opposite views: such scientists are often referred to as scientific dissidents. For example, it is the scientific consensus that HIV causes AIDS but a few scientists disagree. Journalists are advised to go with the scientific consensus.

Avoid false balance

As explained above, it is important to give both sides of a story. However, this approach can go wrong when it is used to give space to views that differ from the scientific consensus on an issue. A good example is climate change. For years the BBC would always ensure that it quoted a climate denier each time it ran an article about climate science, implying the two points of view had equal weight. In 2011, a BBC Trust report (see http://downloads.bbc.co.uk/bbctrust/assets/files/pdf/our_work/science_impartiality/science_impartiality.pdf) concluded that the BBC had fallen victim to "an over-rigid application of the [editorial] guidelines" (BBC Trust, 2011).

Report scientific uncertainty accurately

Uncertainty in science can also generate controversy. You should report accurately. If scientists reveal uncertainty over an issue, report accurately on the contrasting views without overstressing the uncertainty so that it becomes the top issue. Science keeps evolving and as a body of knowledge new studies will always come along that add knowledge to the existing ones. For instance, Albert Einstein's first theory of relativity evoked controversy for a long time until decades of research validated his findings.

Peer review

Scientists announce their research findings by writing a paper and getting it published in a journal. A credible journal will always ask experienced scientists in the same field to scrutinize the paper before it is accepted for publication. This process, referred to as **peer review**, helps to check that the researcher has followed the right procedures and made reasonable interpretations and conclusions in the study. In effect, this protects society from mistaken and fraudulent research findings and recommendations.

In this regard, journalists are advised to report only research findings that have been peer reviewed and published, especially when covering a controversial issue.

Focus on scientific evidence rather than politics

The politicization of certain scientific issues, such as climate change and genetic engineering, means there will often be hot debates and dissenting views on those issues. It is important to focus on the scientific facts, as opposed to the political debate.

Don't take sides

The journalist's job is to accurately give an account of events and facilitate debate. Be inquisitive and fair, in order to bring out a well-researched and balanced story. Don't give your own opinions – especially on controversial issues.

Summary

- Ethics in science journalism is about the acceptable standards of conduct that one must follow in reporting on research.
- Ethical reporting promotes trust between the media and their audiences. It also enhances the credibility of the journalist and their media outlet.
- On the other hand, unethical reporting can cause harm to society and erode public trust in science.
- It is important to be aware of the ethical principles not only of journalism but also of scientific research.
- In summary, ethical principles revolve around truth, maximizing benefit while minimizing harm, fairness and justice.
- Differences of opinion often exist within the scientific community or between scientists and society.
- Journalists have an informational role to play in helping their audiences understand the underlying issues in scientific controversies.
- The decision on whether to report about a controversy should be guided by the public's need to know versus the potential harm to society.
- As a journalist, you should be aware that media coverage of controversial issues in science could have a big impact on both national policymaking and individual decision-making.
- Reporting on scientific controversies can be thrilling but you must ensure that you do it objectively, accurately and responsibly. However controversial the issue may be, your story should do more good than harm to society.

References

BBC Trust (2011) BBC Trust review of impartiality and accuracy of the BBC's coverage of science. BBC Trust July. Available at: http://downloads.bbc.co.uk/bbctrust/assets/files/pdf/our_work/science_impartiality/science_impartiality.pdf (accessed 30 April 2021).

Radford, T. (2008) Reporting on controversies in science. SciDev.net 12 February. Available at: https://www.scidev.net/global/practical-guides/reporting-on-controversies-in-science/ (accessed 30 April 2021).

Resplandy, L., Keeling, R.F., Eddebar, Y., Brooks, M.K., Wang, R., Bopp, L., Long, M.C., et al. (2018) Quantification of ocean heat uptake from changes in atmospheric O_2 and CO_2 composition. Nature 563, 105–108.

Resplandy, L., Keeling, R.F., Eddebar, Y., Brooks, M.K., Wang, R., Bopp, L., Long, M.C., et al. (2019) Retraction note: Quantification of ocean heat uptake from changes in atmospheric O_2 and CO_2 composition. Nature 573, 614.

Sarathchandra, D. and McCright, A. (2017) The effects of media coverage of scientific retractions on risk perceptions. SAGE Open 26 May. Available at: https://journals.sagepub.com/doi/full/10.1177/2158244017709324 (accessed 30 April 2021).

Sibiko, K.W., Veettil, P.C. and Qaim, M. (2018) Small farmers' preferences for weather index insurance: Insights from Kenya. Agriculture & Food Security 7, 53.

SPJ (Society of Professional Journalists) (2014) SPJ Code of Ethics. Available at: https://perma.cc/HAT8-4285 (accessed 30 April 2021).

Simplifying Scientific Facts, Numbers and Statistics

In this chapter
- Simplifying scientific facts
- How to read and understand scientific facts
- Techniques for simplifying scientific facts
- Techniques for simplifying numbers and statistics

Simplifying scientific facts

In communicating science, it is important to ensure the information is simple and clear so that non-specialists can understand it. People can only make meaningful use of information when they understand it clearly.

Learning activity 12.1: Why do non-specialists need to know about science?

Look at the following categories of people. Why would each one of them need to know about new research findings? List the categories and, against each of them, state why they would need to know about research findings.
- Journalists
- Legislators
- Heads of state
- Community members

Please write down your response before reading on.

Feedback
- Journalists need to know about research findings in order to report on them.
- Legislators need to know about research findings in order to inform their debates.
- A head of state needs to know about research findings in order to inform their policy decisions.
- Community members need to know about research findings in order to inform their choices.

Most of these people are not scientists but it is important for them to know about research findings that are relevant to society. It is therefore important to communicate science in a way that is understandable to them. Even if some of these people have studied sciences, they may still struggle to understand the jargon of a different scientific field: for example, a professor of plant breeding may not be familiar with the technical language of nuclear physics.

©2022 CAB International. Science Communication Skills for Journalists: A Resource Book
for Universities in Africa (Ed. Charles Wendo)
DOI: 10.1079/9781789249675.0012

Simple is not simplistic

Some scientists and communicators argue that simplifying scientific information reduces its accuracy and authenticity. This is not true. Simple is different from simplistic. To simplify a message is to make it so clear that the message can be understood easily. Simplistic implies that the message is shallow to the extent that it loses meaning. In the sections below, you will see how to make scientific information easily understandable without undermining its veracity.

How to read and understand scientific facts

One can only simplify what one understands. Before starting to write about a particular piece of research, journalists and communication specialists need to understand the research findings. Therefore, it is important to develop the ability to read and understand scientific literature. There are various things journalists can do to understand unfamiliar words and concepts they might come across when reading a scientific article:

- look up unfamiliar words online immediately
- write down unfamiliar words and look them up later
- refer to relevant background information
- ask the scientist concerned
- read on until the unclear terms become clear.

Techniques for simplifying scientific facts

Below are a number of tried and tested ways for communicating scientific ideas as simply and clearly as possible (summarized in Box 12.1):

- avoid technical jargon where possible
- if you must use technical terms, explain them
- compare with something most people are familiar with
- use relevant real-life examples
- use of visuals and sound (if working in broadcast media)
- avoid information overload
- show your audiences something to see, touch and feel.

We shall explain them one by one.

Avoid technical jargon where possible

Technical jargon makes information difficult to understand and makes it more likely that it will be misinterpreted. It will also put off your audience. Avoid scientific jargon as much as you can. You can almost always say the same thing using alternative, everyday words.

> **Examples**
>
> Use "rainfall"
> Instead of "precipitation"
>
> Use "typhoid germs are rapidly becoming resistant to drugs"
> Instead of "the emergence of multidrug-resistant *Salmonella typhi* is happening at an unprecedented rate"

121

Learning activity 12.2: Say it in simpler words

Read the text below. If you find any unfamiliar words, use the methods suggested above to help you understand.

> Sanushka Naidoo (South Africa) is dedicated to plant defense in the forest species, with an emphasis on Eucalyptus. Her research is focusing on mechanisms that can confer broad-spectrum, long lasting resistance by dissecting gene families and responses to pests and pathogens.
>
> (Next Einstein Forum, 2017)

Write down the main point from the above statement in simpler words.

Please write down your response before reading on.

Feedback

> "Sanushka Naidoo is developing disease-resistant varieties of eucalyptus."

Don't worry if your sentence is a bit different from ours. Writing a layperson's version of a scientific statement is a creative process in which two people are unlikely to come up with identical results.

Learning activity 12.3: Technical versus simpler words and phrases

Match the technical words and phrases in the first column with the most appropriate simpler words and phrases in the second column.

Technical words and phrases	Simpler words and phrases
Aquifers	Medicine effectiveness
Climate-smart crops	Drought-resistant crops
Fibrin-based hydrogel 3D lung cancer model	Underground water
Hepato-cellular carcinoma	Commonest
Hydrocarbon reservoirs	Liver cancer
Medicine efficacy	Artificial lung cancer tissue
Most prevalent	Smell
Olfactory receptor usage	Rainfall
Parasitoid	Under-water soil
Precipitation	Fluid thickness
Subaqueous soil systems	Insect
Viscosity index	Oil and gas

Please write down your response before reading on.

Feedback

Technical words and phrases	Simpler words and phrases
Aquifers	Underground water
Climate-smart crops	Drought-resistant crops
Fibrin-based hydrogel 3D lung cancer model	Artificial lung cancer tissue
Hepato-cellular carcinoma	Liver cancer

Continued

Learning activity 12.3. Continued.

Hydrocarbon reservoirs	Oil and gas
Medicine efficacy	Medicine effectiveness
Most prevalent	Commonest
Olfactory receptor usage	Smell
Parasitoid	Insect
Precipitation	Rainfall
Subaqueous soil systems	Under-water soil
Viscosity index	Fluid thickness

If you must use technical terms, explain them

- "Dental caries, commonly known as tooth decay, is common among school children."
- "His eyes turned yellow, a condition medically referred to as jaundice."
- "Farmers are advised to grow crops that could attract the Fall Armyworm away from maize, a method called trap cropping."
- "Many countries are experiencing high levels of both obesity and undernutrition, a situation referred to as the double burden of malnutrition."
- "The county's average annual rainfall is 1,400 mm, meaning every square metre of ground or roof space receives 1,400 litres of water in a year."

In all of the above examples, the technical terms are used without leaving anybody wondering what they mean.

Below is an extract from a BBC story, "How bacteria are changing your mood" (see %20 https://www.bbc.com/news/health-43815370), written by James Gallagher (2018). The first seven sentences from the story are copied below. Out of 142 words, only two – "microbiome" and "psychobiotics" – are technical terms, and they are both explained.

If anything makes us human it's our minds, thoughts and emotions. And yet a controversial new concept is emerging that claims gut bacteria are an invisible hand altering our brains. Science is piecing together how the trillions of microbes that live on and in all of us – our microbiome – affect our physical health. But even conditions including depression, autism and neurodegenerative disease are now being linked to these tiny creatures.

We've known for centuries that how we feel affects our gut – just think what happens before an exam or a job interview – but now it is being seen as a two-way street. Groups of researchers believe they are on the cusp of a revolution that uses 'mood microbes' or 'psychobiotics' to improve mental health. The study that ignited the whole concept took place at Kyushu University in Japan.

Compare with something most people are familiar with

Compare sizes, shapes, colours, time and so on with something that your audience is familiar with, to help them visualize it. In this case you may not need scientific exactness: instead, use approximations. For example, if you say something has a diameter of 9 inches, not many people will visualize it but if you say it is approximately the size of a standard soccer ball, most people will get it. Distance to a planet can be explained in terms of the number of years it would take a jet plane travelling at its top speed to reach it. You should always try to think about ordinary things that can be compared with the concept you are trying to explain.

Example

In the article "Fossils reveal how ancient birds got their beaks" (see https://www.sciencemag.org/news/2018/05/fossils-reveal-how-ancient-birds-got-their-beaks), published by *Science* magazine, the writer describes an ancient seabird as "having a body like a modern bird, with a snout lined with teeth like a dinosaur" (Vogel, 2018). Many people will be familiar with drawings of a dinosaur, even though they are now extinct, and everyone knows the basic shape of a bird's body.

Use relevant real-life examples

It is important to use real-life examples to help audiences understand what you are explaining. The significance of Ebola can be shown by telling the stories of affected individuals, families and communities. The impact of global warming can be explained through stories of people displaced by flooding that results from it. Technical descriptions can be made clearer by providing some specifics – for example, "high refined carbohydrates, such as white bread, biscuits and sweets".

Learning activity 12.4: Using examples

The extract below is from a WHO article on drug resistance. Read it and identify the sentence where an example is used.

Why is antimicrobial resistance a global concern?

New resistance mechanisms are emerging and spreading globally, threatening our ability to treat common infectious diseases, resulting in prolonged illness, disability, and death.

Without effective antimicrobials for prevention and treatment of infections, medical procedures such as organ transplantation, cancer chemotherapy, diabetes management and major surgery (for example, caesarean sections or hip replacements) become very high risk.

Antimicrobial resistance increases the cost of health care with lengthier stays in hospitals and more intensive care required.

Antimicrobial resistance is putting the gains of the Millennium Development Goals at risk and endangers achievement of the Sustainable Development Goals.

(WHO, n.d.)

Please write down your response before reading on.

Feedback

Without effective antimicrobials for prevention and treatment of infections, medical procedures such as organ transplantation, cancer chemotherapy, diabetes management and major surgery (for example, caesarean sections or hip replacements) become very high risk.

Use of visuals and sound

Graphic illustrations, videos, photos, info videos and drawings are great ways to show what you're trying to explain. They will help your audience quickly appreciate something that would require a lot of words to describe. Similarly, sounds provide the audience with a kind of understanding that cannot be achieved with words.

Avoid information overload

Don't try to present all the information you have gathered about the topic. When learning about a research study, most people simply want to know the findings and the implications of those findings. A detailed description of the methodology and literature review would be of interest to other researchers, especially those in the same field, but not to policymakers and the general public. Providing too much technical detail can turn off your audience. It is important to just focus on what most people want or need to know, and preferably you should stick to your main theme.

Example

When encountering the story "One-dose cholera vaccine gives 90 per cent protection" (see https://www.scidev.net/global/news/one-dose-cholera-vaccine-gives-90-per-cent-protection-1x/) (Maconi, 2018), published by SciDev.net on 13 March 2018, most people will want to know how effective the vaccine is, how soon it might become available to the majority and what it might cost. The scientists involved, on the other hand, might want the article to explain in detail how they designed the vaccine, what scientific principles they used etc, which may not necessarily be of interest to the general public.

Show your audiences something to see, touch and feel

The use of physical objects that people can see, touch and feel – also called artefacts or props – is one of the best ways to help them understand and remember what they are hearing.

Example

If you've been to a dentist you will most likely have seen a model of human jaws with a full set of teeth: dentists use these to explain dental issues to their patients.

Techniques for simplifying numbers and statistics

We interact with numbers every day – for example, managing our expenses, comparing prices, reviewing our children's school report cards and following recipes. However, this doesn't mean we're all comfortable with them. Many readers, listeners and viewers of stories produced by journalists are not good at grasping and applying numbers. Moreover, scientists' research papers include a lot more complicated figures than most of us encounter in our daily lives. It follows that if you report research findings without simplifying the numbers and statistics, your audience might not understand them. This will defeat the purpose of writing the story.

Box 12.1: Ways to simplify science.
- Avoid technical jargon where possible
- If you must use scientific terms, explain them
- Compare with something that most people are familiar with
- Use relevant real-life examples
- Make good use of images and audio
- Avoid information overload
- Use statistics sparingly
- Show your audiences something to see, touch and feel

In a nutshell, the average person is not good with numbers, let alone scientific presentation of statistics. Below are general tips on how to handle numbers and statistics in your journalism:

- using fewer numbers in a sentence
- limiting the number of digits and decimal places
- using familiar fractions to replace percentages
- using infographics
- relating numbers to familiar items.

Using fewer numbers in a sentence

Using too many numbers in a sentence makes it hard for people to read or listen to your story.

Learning activity 12.5: Too many numbers

Read the sentence below aloud.

> The prevalence of HIV changed from 30.4 % in 2011 to 28.1 % in 2012, 24.9 % in 2013, 20.1 % in 2014, 17.8 % in 2015 and 15.3 % in 2016.

Most people would find the above sentence difficult to read. Now, write down a simpler version of the text.

Please write down your response before reading on.

Feedback

> The prevalence of HIV declined steadily from 30 % to 15 % between 2011 and 2016.

If you have an accompanying infographic, you do not necessarily have to state the prevalence figures in the sentence. You could instead say:

> The prevalence of HIV reduced by half between 2011 and 2016.

The lesson is: you can make your sentence easier to understand by reducing – or even eliminating – numbers.

Limit the number of digits and decimal places

Now you know you can make a sentence easier to understand by using fewer numbers. However, a complex number can make your sentence difficult, even if it's the only number in the story. If a number is difficult for you to read aloud, your audience will also find it hard to understand. You can make a sentence more understandable by reducing the number of digits and decimal places.

Example

The sentence, "The world has 7,632,819,325 inhabitants" is difficult to read. "The world has 7.6 billion inhabitants" is easier to read. From this example, it is clear that you can communicate the same message better by using fewer digits.

Think about why you're using numbers in a story. Are you trying to compare two places or groups of people? Are you showing an increase or a decrease over a period? Either way, you probably don't need to use complex numbers. Let's say the population of city A is 20,775,237 while that of city B is 12,240,879. You don't need all those digits to show that city A has a bigger population than city B. You can certainly achieve that objective better by rounding off the population of city A to 20.8 million and that of city B to 12.2 million.

Similarly, numbers with decimal places can be rounded off to the nearest whole number to make them more comprehensible. Suppose the percentage of female scientists was 24.834% in country A and 38.916% in country B: you can round these off to 25% and 39%.

If, on the other hand, you were comparing close figures like 16.851 and 17.237, rounding off to the nearest whole number would bring both figures to 17. In that case you may consider retaining one decimal place, so you end up with 16.9 and 17.2.

When rounding off decimal places, figures from .5 and above are rounded off to the higher number while figures below .5 are rounded off to the lower number. For instance 40.3 is rounded off to 40 while 53.7 becomes 54.

Learning activity 12.6: Rounding off

Round off the numbers in the left column to the nearest whole number.

Original number	Rounded off to
3.445	
19.519	
1009.56	
79.87	
26.483	

Please write down your response before reading on.

Answers

Original number	Rounded off to
3.445	3
19.519	20
1,009.56	1,010
79.87	80
26.483	26

Use familiar fractions

Most people are familiar with fractions like half, a third, a quarter and three-quarters. Rounding off your statistics to the nearest fraction will help your readers' understanding. For example, 33.8% can be referred to as approximately one-third or one in three. We know that one-third is 33.33% and not 33.8% but use of the word "approximately" exonerates you. Below are some of the words and phrases you can use to approximate a percentage to the nearest familiar fraction:

* about
* approximately
* over
* a little more than
* slightly less than
* close to
* nearly.

Learning activity 12.7: Using familiar fractions

Match each of the following percentages with the right phrase, based on a familiar fraction.

Percentage	Phrase based on familiar fraction
26.37%	Nearly half
54.54%	About a quarter
48.73%	More than half
99.39%	Approximately two-thirds
66.91%	Nearly all

Please write down your response before reading on.

Answers

Percentage	Phrase based on familiar fraction
26.37%	About a quarter
54.54%	More than half
48.73%	Nearly half
99.39%	Nearly all
66.91%	Approximately two-thirds

Use infographics

If using familiar fractions makes your message easy to understand, using infographics makes it even easier.

Learning activity 12.8: Pie charts

State the fraction that the shaded area represents in each of the following pie charts.

Continued

Learning activity 12.8. Continued.

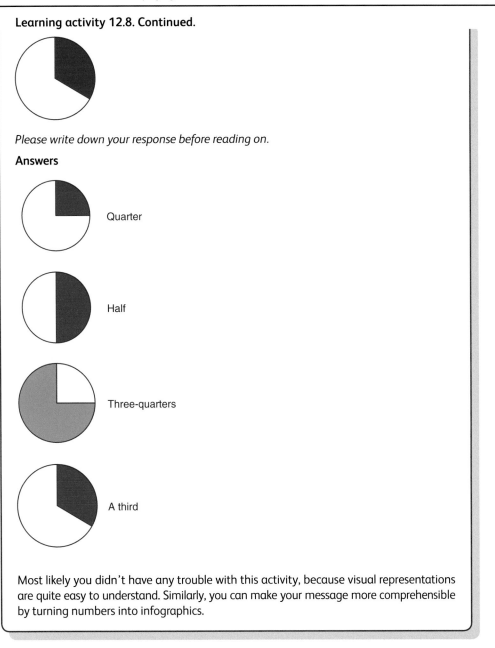

Please write down your response before reading on.

Answers

Quarter

Half

Three-quarters

A third

Most likely you didn't have any trouble with this activity, because visual representations are quite easy to understand. Similarly, you can make your message more comprehensible by turning numbers into infographics.

Learning activity 12.9: Counting without counting
Take a quick look at the image below and answer the question without counting the stars.
Which group has more stars, A or B?

Continued

Learning activity 12.9. Continued.

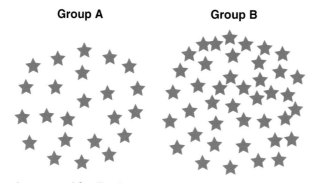

Group A   **Group B**

Answer and feedback

Group B has more stars. Human beings are born with the ability to estimate quantities at a glance without counting. You can help your audience to quickly make sense of numbers by using infographics.

Generating infographics

An infographic brings together graphs, numbers, text, drawings and photos to explain an idea. Some infographics may have all of these elements: for example, some have only text and graphs, text and numbers, or text and drawings. An infographic is normally visually attractive and it usually communicates a lot at a glance. You can work with a graphic designer to produce an infographic, or learn how to make one yourself. Below are some online tools that you can use to generate infographics.

- Piktochart (see https://piktochart.com/)
- Canva (see https://www.canva.com/ff_sn/)
- Infogram (see https://infogram.com/)
- Google Sheets (see https://www.google.com/sheets/about/)

Note that technology changes at a fast rate so by the time you read this book newer tools might be available. As a journalist, it is important to keep abreast of new technologies.

Compare numbers with familiar items

Numbers can create confusion if they are simply thrown at the audience. For example, many people may not be able to easily visualize how high 70 centimetres is, but they will most likely be familiar with the height of a standard classroom desk.

Below are some examples of measurements and familiar items.

Measurements	Familiar items
50,000 square feet	About the area of a soccer pitch
1,000 km per hour	Faster than a jet plane
25 inches long	Arm's length
Nine inches in diameter	The size of a soccer ball
60,000 square kilometres	About the size of Lake Victoria
40 inches high	Taller than an office desk
Eight inches in length	The length of an average table fork

Summary

To make scientific information understandable to non-scientists, you need to:

- avoid technical jargon where possible
- explain scientific terms, where you must use them
- compare with something that most people are familiar with
- use relevant real-life examples
- use good photos, infographics and audio (if applicable)
- avoid information overload
- give your audiences something to see, touch and feel.

You can also simplify numbers and statistics by:
- using fewer numbers in a sentence
- limiting the number of digits and decimal places
- using familiar fractions
- using infographics
- relating numbers to familiar items.

References

Gallagher, J. (2018) How bacteria are changing your mood. BBC.com 24 April. Available at: https://www.bbc.com/news/health-43815370 (accessed 1 May 2021).

Maconi, M. (2018) One-dose cholera vaccine gives 90 per cent protection. SciDev.net 14 March. Available at: https://www.scidev.net/global/news/one-dose-cholera-vaccine-gives-90-per-cent-protection-1x/ (accessed 1 May).

Next Einstein Forum (2017) Meet the 2017–2019 NEF Fellows. Available at: https://nef.org/announcing-2017-2019-next-einstein-forum-fellows-africas-top-scientists-solving-global-challenges/ (accessed 17 February 2022).

Vogel, G. (2018) Fossils reveal how ancient birds got their beaks. Sciencemag.org 2 May. Available at: https://www.sciencemag.org/news/2018/05/fossils-reveal-how-ancient-birds-got-their-beaks) (accessed 1 May 2021).

WHO (n.d.) About AMR. WHO Regional Office for Europe. Available at: https://www.euro.who.int/en/health-topics/disease-prevention/antimicrobial-resistance/about-amr (accessed 1 May 2021).

Packaging a Science Story

In this chapter
- What makes a good science story?
- How to structure a science story

Editors want a science story that will be interesting for, and important to, their readers, viewers and listeners. Such a story should be engaging in terms of both its content and how it's structured. This chapter is about the ingredients of a good science story – and how to package the story.

What makes a good science story?

Competition for space in the media, and for people's attention, is getting tougher. When faced with more stories than they can publish, an editor will select only the best of the newsworthy stories that they receive. Similarly, readers, viewers and listeners have a choice: if they don't find your story interesting, they can ignore it. Therefore, it is not enough to write a story that just meets the basic criteria for newsworthiness: you need to give editors and audiences a reason to prefer your story over others.

Learning activity 13.1: The qualities of a good science story
In your view, what are the essential qualities of a good science story?

Please write down your response before reading on.

Feedback
A good science story should:
- be of interest to non-specialists
- be easy to understand
- be accurate
- have information and views from various sources
- go beyond official information
- help audiences understand an issue better.

To better understand the qualities of a good story, let's discuss them in relation to a particular research study.

©2022 CAB International. Science Communication Skills for Journalists: A Resource Book for Universities in Africa (Ed. Charles Wendo)
DOI:10.1079/9781789249675.0013

In Tanzania, a study by Maleko and Koipapi (2015, see http://www.lrrd.org/lrrd27/4/male27070. html) concluded that a shortage of pasture is a serious problem in Longido district, especially during the dry season from August to November. During these months, livestock productivity declines due to scarcity of pasture and water resources. As a result, herders move longer distances and spend more hours grazing and watering their livestock. The main causes of pasture shortage are prolonged droughts and increased cultivation, which reduce the size of land available for grazing.

Learning activity 13.2: A good science story is interesting to non-specialists

The issue of pasture shortage in Longido district could potentially be seen as uninteresting to those who are not directly affected. How would you make the pasture study described above interesting to a wider audience?

See Chapter 14, "Bringing a Science Story to Life", for more on how to make science interesting to non-specialists.

Please write down your response before reading on.

Feedback

- Humanize the story by narrating the real-life experiences of affected herders and experts involved with the problem.
- Present facts, figures and experiences to show how a similar pasture shortage affects other parts of the country and continent.
- Show why people in other parts of the country should be concerned about the livestock feed shortage in Longido district: for example, does it affect the availability and price of milk and meat in other districts?
- Use a captivating picture that depicts the shortage of pasture and water.
- Include a discussion of the policy implications of the research findings.

Learning activity 13.3: A good science story is easy to understand

Scientists communicate their research findings in a format that is difficult for non-specialists to understand. What would you do to simplify the information from the pasture study discussed above so that non-specialists can understand it? (If you've read Chapter 12, you should already have some good ideas about this.)

Please write down your response before reading on.

Feedback

- Avoid technical jargon. Replace technical words with simpler words and expressions: for example, "the main causes of pasture shortage" instead of "major drivers of pasture inadequacy".
- If you must use technical terms, explain them.
- To avoid information overload, narrow the story down to the most important research finding and its implications.
- Use descriptions that help the reader visualize the issue: for example, saying the pasture has been grazed down to bare soil.
- Use visuals, such as infographics, photos and videos (if applicable).
- Organize the story logically: for example, begin by narrating the herders' experiences (we discuss how to structure a science story later in this chapter).

Learning activity 13.4: A good science story is one that is accurate

In reporting science, it is important to get it right, otherwise you can mislead your audience, which can have serious consequences. What would you do to ensure the accuracy and credibility of your story on the pasture study discussed above?

Please write down your response before reading on.

Feedback

- Check that the journal where the study was published is credible – for example, by looking it up on Journal Citation Reports (see https://clarivate.com/webofsciencegroup/solutions/journal-citation-reports).
- Read the research paper, as opposed to relying on a press release.
- Check for the meanings of technical expressions before paraphrasing them.
- If in doubt, ask the scientist to confirm that you have paraphrased correctly.
- Interpret the research findings in the right context .
- Add a comment from an independent expert working in the same field.

Learning activity 13.5: A good science story has information and views from multiple sources

Having a diversity of voices makes a story more informative and balanced. Before reading on, list the categories of people you would interview for a story on the pasture study discussed above.

Please write down your response before reading on.

Feedback

Here are some suggestions of the voices you could include in the story.

- Herders – on how the pasture shortage is affecting their livelihoods.
- The researcher – on their findings and their implications.
- Consumers of livestock products – on any changes in their access to, and the affordability of, livestock products.
- Experts in similar fields who did not participate in the research (experts on rangelands, pasture, livestock feeding) – for independent views on the implications of the research findings.
- A policymaker or implementer – to understand the policy and programme implications of the research findings.

Learning activity 13.6: A good science story goes beyond official information

Sometimes the facts on the ground are different from the impression that is given by government officials. It is important to independently investigate information, rather than just relying on the official government story.

Regarding the pasture shortage study discussed above, how would you go about verifying the information given by government officials?

Please write down your response before reading on.

Feedback

- Travel to the affected area, make observations and speak to the herders.
- Read independent reports, if there are any.

Learning activity 13.7: A good science story helps audiences understand an issue better

A good science story goes beyond the presentation of facts to help people make sense of what is happening. Regarding the pasture shortage study discussed above, what kind of information would you include to give your audiences a deeper understanding of the pasture shortage?

Please write down your response before reading on.

Feedback

- What is causing the pasture shortage?
- Are there any historical, social and environmental factors that need to be addressed?
- Why does this area suffer prolonged droughts?
- Why has there been a decrease in the area available for grazing? For example, is it related to population growth? Are there any changes in land policies or tenure systems?
- To what extent might this pasture shortage affect the availability of livestock products, such as milk and meat?
- What can be done at the policy level?
- At the personal level, what can herders do, and who can help them to achieve that?
- Are there success stories, where communities elsewhere have overcome a similar pasture shortage?

How to structure a science story

Having looked at the ingredients of a good science story, let's turn to how we can package the story.

Activity 13.8: Two different structures

Take a look at the following two ways of beginning a story and spot the difference.

Story 1

Headline

"Banana plants can be primed to resist warm weather"(see https://www.scidev.net/asia-pacific/news/banana-plants-can-be-primed-to-resist-warm-weather/)

First paragraph (also called the "lede")

> A study carried out by the Indian Institute of Horticultural Research (IIHR), Bangalore, suggests that banana plants can be made resistant to hot climatic conditions by subjecting the seedlings to mild heat stress.
>
> (Bhattacharya, 2018)

Story 2

Headline

"'I was told I'd die if I had a baby'" (see https://www.bbc.com/news/health-46123097).

First paragraph

> Hayley Martin, 47, vividly remembers the morning her life changed forever. 'I woke up and I felt very, very poorly. I put my hands to my head and I was drenched in sweat. I knew straight away it was a heart attack,' she told the BBC's Victoria Derbyshire programme.
>
> (Melley, 2018)

Continued

Activity 13.8. Continued.

What differences do you notice between the first paragraphs of the two stories?

Please write down your response before reading on.

Feedback

	Lede of story 1	Lede of story 2
Length	Shorter	Longer
Content	Summarizes the story	Gives a clue but doesn't reveal the whole story
Style	Goes straight to the facts	Tells a story
Proximity	Has no actual person	Has an actual person

A good story is not simply a collection of facts: a story should have a recognizable structure and a logical flow. There are several ways to structure a science story. Story 1 above has an inverted pyramid structure, with the most important facts summarized in the first sentence. Story 2 has a different structure: it begins with an anecdote, followed by key facts and discussions. This is referred to as the *Wall Street Journal* structure. The inverted pyramid and the *Wall Street Journal* structure are among the most used story structures in journalism. We shall look at these two structures in more detail below.

Having a structure in mind for your story helps you plan better by indicating what information you will need to gather for your article. In particular, it helps you to know what questions to ask when interviewing sources and what to look out for when examining documents or data. Most importantly, it helps you to sketch out the story before you start writing.

The inverted pyramid structure

The inverted pyramid is the most used, and the oldest, structure in news writing. It begins with the most important facts, followed by the next most important and so on.

Lede/lead/intro

In inverted pyramid writing, the first paragraph – also called the lede or intro – is a concise statement highlighting the most important facts of the story. Most often, this is a one-sentence paragraph. It should answer as many of the "5Ws+H" as possible (who, what, when, where, why and how). If possible it should also answer the question "so what?".

- What is the most important research finding?
- Who carried out the study?
- When were the findings published?
- Where (in which journal) were the results published? Where was the study carried out?
- Why was the research carried out?
- How was the study carried out?

Learning activity 13.9: Inverted pyramid structure lede

Take a look at the lede below and identify the "5Ws".

> WASHINGTON (Reuters): Using a radar instrument on an orbiting spacecraft, scientists have spotted what they said on Wednesday [25 July 2018] appears to be a sizable salt-laden lake under ice on the southern polar plain of Mars, a body of water they called a possible habitat for microbial life."

> (Dunham, 2018, see https://www.reuters.com/article/space-mars-idINKBN1KF1ZL)

Continued

Activity 13.9. Continued.

Please write down your response before reading on.

Feedback

- Who – A scientist, Roberto Orosei of Istituto Nazionale di Astrofisica, in Italy. His name is not in the lede but he is quoted in subsequent paragraphs.
- What – They discovered a hidden lake that might be home to living microorganisms.
- Where – On Mars
- When – They made the announcement on Wednesday 25 July 2018.
- How – Using a radar instrument on an orbiting spacecraft.

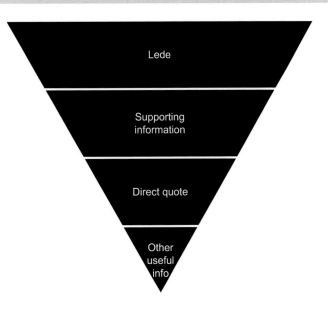

Fig. 13.1. Typical inverted pyramid structure of a science news story.

Immediately after the lede you should add the next most important information, in a few sentences. This is information that sheds more light on what was revealed in the lede. In the case of the hidden lake on Mars, the supporting sentences are as follows:

> The reservoir they detected – roughly 12 miles (20 km) in diameter, shaped like a rounded triangle and located about a mile (1.5 km) beneath the ice surface – represents the first stable body of liquid water ever found on Mars. Whether anywhere other than Earth has harbored life is one of the supreme questions in science, and the new findings offer tantalizing evidence, though no proof. Water is considered a fundamental ingredient for life. The researchers said it could take years to verify whether something is actually living in this body of water that resembles a subglacial lake on Earth, perhaps with a future mission drilling through the ice to sample the water below.
>
> (Dunham, 2018)

Direct quote

After the supporting information, you need a good direct quote to colour your story and place more emphasis on a few areas. In our example the first direct quote is the following:

> **Learning activity 13.10: *Wall Street Journal* structure lede**
>
> Look at the lede below in science writer Jane Qiu's feature article, "Trouble in Tibet", published in *Nature* (see https://www.nature.com/news/trouble-in-tibet-1.19139).
>
>> In the northern reaches of the Tibetan Plateau, dozens of yaks graze on grasslands that look like a threadbare carpet. The pasture has been munched down to bare soil in places, and deep cracks run across the snow-dusted landscape. The animals' owner, a herder named Dodra, emerges from his home wearing a black robe, a cowboy hat and a gentle smile tinged with worry.
>>
>> (Qiu, 2016)
>
> What elements of this lede suggest it follows the *Wall Street Journal* structure?
>
> *Please write down your response before reading on.*
>
> **Feedback**
>
> It begins with a moving, descriptive story of a real human being, without disclosing what the main point of the story is in the first paragraph.

"This is the place on Mars where you have something that most resembles a habitat, a place where life could subsist," said planetary scientist Roberto Orosei of Istituto Nazionale di Astrofisica in Italy, who led the research published in the journal *Science*.

(Dunham, 2018)

See Chapter 14, "Bringing a Science Story to Life", for more information on identifying and using the "golden quote".

Any other additional information and quotes

After the direct quote you can include any other information and quotes that will add substance to the story. This is where you would include comments from other sources.

The description in the paragraphs above is a recommended flow of an inverted pyramid story: in practice, you may find it necessary to adjust the structure (for example, a direct quote could come earlier than suggested here).

Wall Street Journal structure

This structure is recommended for feature articles, though it may also be used for news. It begins with an enticing short story, quote or description to pull people into the story without necessarily disclosing what the story is all about at the beginning.

Whatever you write in the lede must be attention-grabbing and concise. One way of doing this is to think about all the information you have gathered. Think about everything you heard from interviewees, observed or read. Make a list of everything you found interesting, surprising, impressive, exciting, alarming or catchy about the individuals associated with the topic you are writing about. Decide which of these stands out and then use it to construct a lede.

Nut graf

Still thinking about our story concerning the pasture shortage in Longido district: having drawn people in using an anecdote, you need to move on to the purpose of your story. You are not simply entertaining them with a colourful story of herders: you need a sentence or paragraph that summarizes the main point of the story. That sentence or paragraph(s) is called a **nut graf** (short for "nutshell paragraph"). For example, in the "Trouble in Tibet" story referred to above, the nut graf comes in the fifth and sixth paragraphs:

> The challenges that face Dodra and other Tibetan herders are at odds with glowing reports from Chinese state media about the health of Tibetan grasslands – an area of 1.5 million square kilometres – and the experiences of the millions of nomads there. Since the 1990s, the government has carried out a series of policies that moved once-mobile herders into settlements and sharply limited livestock grazing. According to the official account, these policies have helped to restore the grasslands and to improve standards of living for the nomads.
>
> But many researchers argue that available evidence shows the opposite: that the policies are harming the environment and the herders …
>
> (Qiu, 2016)

That is the main point of the story.

Main body of the article

With regard to the main point of the story, what relevant details will your audience want or need to know? That is the information that should be in the main body of the article.

The main body is usually the longest part of the article. This is where you bring in the necessary context, facts, figures, interpretations, arguments, experiences, examples and comparisons. In the case of "Trouble in Tibet" the author traces the history of the Tibetan grazing dilemma and explores the link between government policy and overgrazing. She weaves in various research findings, and explains how the plateau's environmental problems might affect the quality and availability of water in neighbouring countries and aggravate global warming. In building the main body of the story, it is important to stick to the theme and underlying purpose of your story.

Conclusion

Once you've constructed the lede, nut graf and the main body of your article, don't simply stop writing. End the story with a concluding paragraph that helps your readers, listeners and viewers to understand what is going on and to make informed predictions about the future. In that regard it is necessary to refocus them on the main point of the story, summarize the main facts and arguments, and conclude with a projection into the future. In the *Wall Street Journal* structure the conclusion refocuses on the individual whose story was told in the lede. For example, the final sentence in the story could be about what is likely to happen to the person whose story was told in the lede.

Activity 13.11: *Wall Street Journal* **style conclusion**

Take a look at the conclusion below in the article "Trouble in Tibet" and identify at least one element that suggests the article is written following the *Wall Street Journal* structure.

> Far away from Lhasa, herders such as Dodra say that they are not seeing the benefits of government policies. After we finish our visit, at his home, Dodra's entire family walks us into the courtyard – his mother-in-law spinning a prayer wheel and his children trailing behind. It has stopped snowing, and the sky has turned a crystal-clear, cobalt blue. "The land has served us well for generations," says Dodra as he looks uneasily over his pasture. "Now things are falling apart – but we don't get a say about how best to safeguard our land and future."
>
> (Qiu, 2016)

Please write down your response before reading on.

Feedback

The article concludes with a quote from Dodra, the herder who was mentioned in the lede earlier.

Summary

- A good science story is interesting to non-specialists.
- A good science story is easy for non-specialists to understand.
- A good science story is accurate.
- A good science story has information and views from multiple sources.
- A good science story goes beyond official information.
- A good science story helps audiences understand an issue better.
- There are various ways to structure a science story, the most common being the inverted pyramid and the *Wall Street Journal* structures.
- Identifying a structure and sticking to it helps you to write a better story.
- In the inverted pyramid structure, the most important aspects of the story are summarized in the lede – supporting facts, figures and views follow the lede in order of importance.
- The *Wall Street Journal* structure begins with an engaging anecdote and introduces the main point of the story before bringing in context, facts, figures, interpretations, arguments, experiences, examples and comparisons.

References

Bhattacharya, P. (2018) Banana plants can be primed to resist warm weather. SciDev.net 24 July. Available at: https://www.scidev.net/asia-pacific/news/banana-plants-can-be-primed-to-resist-warm-weather/ (accessed 1 May 2021).

Dunham, W. (2018) Underground lake found on Mars, raising possibility of life. Reuters.ecom 25 July. Available at: https://www.reuters.com/article/space-mars-idINKBN1KF1ZL (accessed 1 May 2021).

Maleko, D.D. and Koipapi, M.L. (2015) Opportunities and constraints for overcoming dry season livestock feed shortages in communal semi-arid rangelands of Northern Tanzania: A case of Longido District. *Livestock Research for Rural Development* 27.

Melley, J. (2018) "I was told I'd die if I had a baby". BBC.com, 9 November. Available at: https://www.bbc.com/news/health-46123097 (accessed 1 May e2021).

Qiu, J. (2016) Trouble in Tibet: Rapid changes in Tibetan grasslands are threatening Asia's main water supply and the livelihood of nomads. *Nature* 529. Available at: https://www.nature.com/news/trouble-in-tibet-1.19139 (accessed 1 May 2021).

Bringing a Science Story to Life

14

In this chapter
- Making a science story interesting to non-specialists
- Balancing human interest and science
- Humanizing the science
- Relating your story to a trending topic
- Relating your story to people's needs and interests
- Building blocks of a human interest science story
- A human interest story is not a full life profile
- Whose story do you tell?
- Identifying and using the "golden quote" to colour a science story

Making a science story interesting to non-specialists

Science often involves words and concepts that are complicated and potentially uninteresting to non-specialists. You can bring a science story to life by applying certain known techniques for making science interesting, by telling human interest stories and by identifying great direct quotes.

When an editor says a piece of information is newsworthy, they mean it is interesting, important and timely enough to draw most people's attention, and therefore worthy of being reported on in the mass media. The criteria that editors use for judging the newsworthiness of a story are called **news values** or **news criteria**. The news values apply to all fields of journalism, including science reporting.

Learning activity 14.1: News values
What are the values you would consider in determining what is newsworthy?
Please write down your response before reading on.

Feedback

Below are some of the main news values.
- **Impact** – The number of people a story affects, and the extent to which it affects them.
- **Controversy** – Disagreements within the scientific community or between scientists and other actors, such as politicians.
- **Novelty** – Something new and different, that has not been seen or heard about before.

Continued

©2022 CAB International. Science Communication Skills for Journalists: A Resource Book for Universities in Africa (Ed. Charles Wendo)
DOI:10.1079/9781789249675.0014

Learning activity 14.1. Continued.

- **Surprise or the unusual** – Something amazing or totally unexpected.
- **Prominence** – Related to a famous person or institution or group.
- **Topicality** (also referred to as **timeliness** or **currency**) – Related to something that is on people's minds during a given period.
- **Proximity** – Related to a situation, place, person, institution or situation that audiences are familiar with.
- **Usefulness** – Information that people can put to use in their everyday lives.

Learning activity 14.2: News values

What is the most likely news value of a newly published research study on rangeland degradation in Tanzania?

Please write down your response before reading on.

Feedback

Novelty, because the study is newly published. The other news values may or may not apply depending on where you are and what else is happening around you.

Balancing human interest and science

Whereas it is important to know about news values, not every newsworthy story will get published or be aired. Most editors receive more newsworthy stories and ideas than they can publish/broadcast, so they have to prioritize. Thus you need to go the extra mile to make your story more interesting. This is more important when dealing with a science topic, which can potentially be unexciting to most people depending on how it's presented.

A journalist can do one or more of the following to make a science story interesting to the general public, without getting involved in sensationalism:

- humanize the science
- relate it to a trending topic
- relate it to people's most pressing needs and interests.

Learning activity 14.3: Which is more interesting?

Read the two headlines.

Headline 1: "Drug resistance could make 28 million people poor"
Headline 2: "New genes for drug resistance discovered"

Which would most people click on first, and why?

Please write down your response before reading on.

Feedback

Headline 1: "Drug resistance could make 28 million people poor"

Headline 1 is more relatable to most people. It is about something that matters to most people. It focuses on how antibiotic resistance could affect people, and therefore gives

Continued

Learning activity 14.3. Continued.

them a reason to be concerned about it. But if you selected headline 2, it doesn't mean you are wrong; there are reasons why some people might make a different choice here. The point is that people have choices: they choose what to read, view or listen to depending on various factors, such as where they live, what they do and what their needs and aspirations are. They can choose to ignore your information if they do not relate to it. Most people will read a science story because they can find interesting and useful information in it that relates to their lives, and not simply because they want to know about the latest scientific developments.

Humanizing the science

Telling a remarkable and informative story about a person or people can make scientific information interesting. The person could be a hero, villain, beneficiary or victim vis-à-vis the issue in question.

Learning activity 14.4: Which is more catchy?

Take a look at these two statements and select the one that is more likely to catch an editor's attention.

Statement 1: "Scientists have invented a special implant to restore connection between the brain and an injured spinal cord"

Statement 2: "A paralysed man has been able to walk again after a ground-breaking surgical operation"

Please write down your response before reading on.

Feedback

Statement 2: "A paralysed man has been able to walk again after a ground-breaking surgical operation."

Both statements are derived from the same set of facts. The scientists concerned created a device that helped to bypass a spinal injury and keep the nerves in the legs connected to the brain. As a result, they helped a paralysed man to walk again (more or less). It is the story of a paralysed man walking again, rather than the technical information about the spinal bypass operation, that will make the editor sit up and listen. Of course, the story would not be complete without explaining how the paralysed man has been able to walk again, but it is the human story that will draw the editor's attention. In this case the human story is used as a hook to make people interested in the science.

What if you can't find human beings whose stories can be told? You could still achieve human interest by showing how a scientific issue impacts on human beings. This story offers one example: "Drug resistance could make 28 million people poor" (see https://www.scidev.net/sub-saharan-africa/news/drug-resistance-could-make-28-million-people-poor/) (Ogada, 2019).

Relating your story to a trending topic

Human beings tend to pay attention to the **trending topic** of the moment. Is there an issue that is dominating the newspaper headlines, radio and TV discussions, social media talk, religious sermons or even sitting-room discussions in your society at a given point in time? A story that provides scientific information about a trending topic will attract attention. People like to listen to additional information about an issue that is on their minds at the moment.

How do you know what topic is on people's minds? You need to monitor social media posts, read newspapers, listen to the radio, watch TV and listen to conversations around you, for instance in the workplace, at home and when on public transport. People will most likely pay attention to new and useful scientific information about a topic that is trending on social media or that has recently been reported about in the mainstream media.

> **Learning activity 14.5: A trending topic**
> This activity requires you to have access to the internet.
>
> Look at the front page of the three leading news websites in your country. Look at the headlines to identify the most dominant topic across the three news websites. Now, read one of the articles. Beyond what the journalist reported in the article, what more would you like to know about the issue? What questions come to your mind after reading the article?
>
> *Please write down your response before reading on.*
>
> **Feedback**
>
> Virtually every story in the media will leave you with some unanswered questions. No story will provide all the information that there is to know about the topic. It is almost always possible to provide new and interesting scientific information that informs discussions on a trending topic.

Relating your story to people's needs and interests

You can draw more attention to scientific information by showing how it relates to people's most pressing needs and interests. This is because an issue that ranks high on the list of people's most pressing concerns will always be on their minds even if it is not a trending topic at the moment. For example, if a community perpetually faces hunger, they will pay attention to a story about food. Likewise, information about a new technology that dramatically saves energy will attract wide attention in a society where electricity is expensive and in limited supply.

Writing a human interest science story

In the discussion above, we looked at humanizing science as one of the techniques for making it interesting to media audiences. There are a number of other arguments for engaging in human interest science journalism.

- It is engaging and gets people talking: people like to hear and talk about people, therefore human interest stories will be discussed more frequently than abstract stories.
- Human interest provides a fresh angle to an old science story. Someone is likely to read a human interest story even if they have read other stories on the same issue. Every individual has a unique story.

- It helps your audience to understand and appreciate the story better. For example, explaining how someone has lived with HIV for over 20 years leaves a stronger impression on the reader than just saying antiretroviral treatment works.
- It provides a break from "big man journalism". The individuals reported about in human interest stories need not be famous people: they simply need to have an interesting story to tell. This gives us all a break from the routine focus on prominent scientists, government officials and academics.

Building blocks of a human interest science story

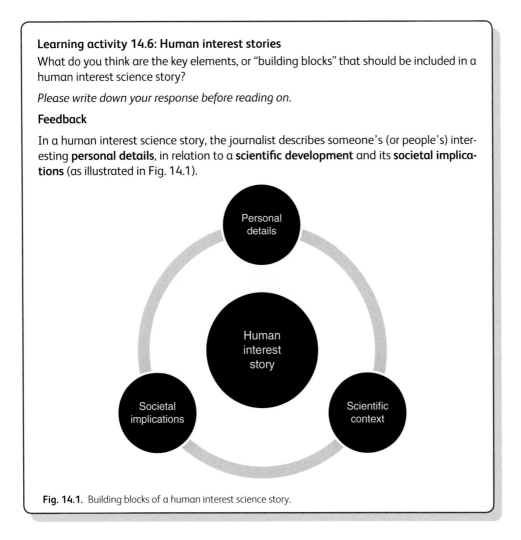

Learning activity 14.6: Human interest stories

What do you think are the key elements, or "building blocks" that should be included in a human interest science story?

Please write down your response before reading on.

Feedback

In a human interest science story, the journalist describes someone's (or people's) interesting **personal details**, in relation to a **scientific development** and its **societal implications** (as illustrated in Fig. 14.1).

Personal details

Human interest story

Societal implications

Scientific context

Fig. 14.1. Building blocks of a human interest science story.

The story "How has Magic Johnson survived 20 years with HIV?" (see https://www.livescience.com/16909-magic-johnson-hiv-aids-anniversary.html) describes how the basketball star bounced back from AIDS. The story begins and ends with Johnson. His name is in the headline, in the first paragraph, in every one to two paragraphs, and in the conclusion. Notably, the article includes a photo of him in action. So the story is about him. But the story is also about people living with HIV. While Johnson is present throughout the story, it is not all about him. The writer relates Johnson's story to the wider community of people living with HIV. He makes it clear that Johnson is one of many people whose lives have been turned around by antiretroviral medication. Look at the paragraph below, extracted from the story:

> The answer to Johnson's survival is far from "magic." According to reports, he takes the same kinds of drugs that are available to other HIV patients in the developed world, and increasingly in impoverished nations in Africa and Asia, where the disease still runs rampant. Many people have lived with HIV even longer than Johnson.
>
> (Hadhazy, n.d.)

The story is also about HIV. The virus is mentioned in almost every paragraph. Like Johnson, HIV features in the headline, and in the opening and concluding paragraphs.

And, finally, the story is about a major scientific discovery that turned around the life of the celebrity, and that has turned around the lives of many other persons living with HIV worldwide. That discovery is the highly active antiretroviral therapy (HAART), often referred to simply as antiretroviral therapy. The writer uses simple language to explain how HIV affects human beings, and what antiretroviral therapy does. So this is a science story. In a nutshell, the writer combines human interest and science to tell an interesting and informative story about a scientific discovery related to HIV treatment.

A human interest story is not a full life profile

Johnson was an internationally renowned basketball star and there are many things one would want to know about his career, but basketball is hardly mentioned in the article about him. The writer of the article doesn't attempt to narrate everything interesting about Johnson: he focuses on those facts that are related to HIV and antiretroviral therapy. Likewise, in any human interest science story, you need to focus only on the personal details that help illustrate the science. In the article, Johnson and HIV are mentioned 19 and 30 times, respectively. Thus, HIV is mentioned more than Johnson. What does this tell us? While a human interest science story needs to include relevant personal information about the person concerned, it should remain focused on the science.

Whose story do you tell?

In science-related developments there can be heroes, villains, victims, survivors, unique beneficiaries and interesting individuals whose stories can be told. Likewise, in many human interest stories there are related scientific explanations that can be provided to your audience. Take for instance the man who managed to retrieve a huge piece of maize meal as city authorities were demolishing an illegal restaurant in Nairobi, Kenya, in August 2018. Journalists followed him up for a human interest story: it turned out he was a homeless person living on the streets in one of the city's suburbs. But, is there a science angle to that human interest story?

Learning activity 14.7: A science angle to a human interest story

Suggest a science angle to the above human interest story of the man who retrieved a piece of maize meal from a restaurant as it was being demolished.

Please write down your response before reading on.

Feedback

To most journalists, this was simply a story of a hungry homeless man. But one science journalist looked at the story through a difference lens. In his Facebook post, he connected the incident to the issue of food wastage amidst frequent hunger. He posted the man's picture alongside infographics on food wastage.

A Facebook post on a man who retrieved a lump of maize meal as city authorities in Nairobi demolished an illegal food kiosk.

A food wastage angle to the human interest story highlighted in Learning activity 14.7 can be interesting and informative. In this regard, the following points can be considered.

- There is a lot of related scientific research that can reinforce the story. A simple search on Google Scholar yields thousands of research papers related to food waste amidst hunger. Many of the findings have not been reported about in the media.
- Food wastage is a growing international concern – the amount of food that goes to waste globally could feed all the hungry people worldwide.
- Plenty of relevant facts and figures have been put together by the Food and Agriculture Organization of the United Nations (FAO) and other institutions, which can help tell the story of food wastage amidst hunger.
- There is an interesting human story to tell. The man with the maize meal was talked about widely on social media for weeks. During that period, the story would attract a lot of attention.

Identifying and using the "golden quote" to colour a science story

We have looked at techniques for making science interesting to non-specialists. We also looked at how to report a good human interest story. Another way to bring a science story to life is to use captivating direct quotes.

When you use the exact words that someone has spoken and put them inside quotation marks you are making a statement that that piece of information deserves special attention. Additionally, you want your audience to know this quote has come from that particular person, in their own words. This means you should select the text that you use as a direct quote in your story carefully: you don't want someone to come to the text understanding that it is being given special attention, only to find a flat statement. A direct quote should be memorable and should illuminate the main point of the person's statement.

Learning activity 14.8: Qualities of a good direct quote

Look at the quote below from the SciDev.Net story "Kenyan farmers suffer losses as climate change grips" (see https://www.scidev.net/sub-saharan-africa/feature/kenyan-farmers-suffer-losses-as-climate-change-grips/). The quote is in reference to a locust invasion.

"They came like dust, wind, and fire in April last year."

(Lillian, 2022)

A good quote is catchy and memorable, and often conveys strong opinions or feelings, rather than plain facts. Which of the following features do you see in the quote above?

- Catchy
- Memorable
- Expresses strong feelings or opinions, rather than plain facts

Please write down your response before reading on.

Feedback

The quote is catchy and memorable.

Learning activity 14.9: Find a direct quote

Look at the excerpts below from a speech by the WHO Director General, Dr Tedros Adhanom Ghebreyesus (see https://www.who.int/director-general/speeches/detail/opening-speech-at-aids-2018), at the opening of the 2018 International AIDS Conference, and identify one paragraph or sentence that can make a good direct quote.

In May this year, the World Health Assembly approved a new 5-year strategy for WHO, to support countries on their journey towards the Sustainable Development Goals.

At its heart are what we call the "triple billion" targets for 2023:

1 billion more people benefitting from universal health coverage,
1 billion more people better protected from health emergencies;
and 1 billion more people enjoying better health and well-being …

Continued

Learning activity 14.9. Continued.

… the reality is that more than half of the world's population lacks access to essential health services, including vaccination, treatment for HIV, hepatitis and TB, family planning services, and the ability to see a health worker.

And every year, almost 100 million people are pushed into extreme poverty by the costs of paying for care out of their own pockets.

This outrage must end.
No one should get sick and die just because they're poor or marginalized…
From the above excerpt, identify at least one sentence that would make a good quote.

Please write down your response before reading on.

Feedback

"This outrage must end" and "No one should get sick and die just because they're poor or marginalized". Each of these can stand alone as a direct quote. They both catch the reader or listener's attention by conveying a strong opinion on what the speaker is discussing.

Learning activity 14.10: What if there is no catchy statement?

Suppose you listened to an entire speech and didn't find a catchy paragraph or sentence to use as a direct quote – which of the following options would be the most appropriate?

- Select the best of the uninteresting sentences
- Do without a direct quote
- Edit one sentence to make it interesting

Please write down your response before reading on.

Feedback

It is better to do without a direct quote than to use a complicated or obvious statement.

It is not right to change someone's words: you should not edit a sentence to create an interesting direct quote. It is better to do without a direct quote than to doctor one. However, you may ask the source additional questions to prompt them to say something that can make a good direct quote. You can ask the same thing in different words. Provocative questions can often yield quotable answers. Ask "Why?" and "How?" as often as possible. If you are reporting on a speech at a formal ceremony, you can follow it up with an interview to get quotable comments.

Learning activity 14.11: A legitimate way to turn technical jargon into a golden quote

We have already seen that it is not right to alter someone's words. However, there are ways to increase the chances of your source giving you a "golden quote". Can you think of one?

Please write down your response before reading on.

Continued

Learning activity 14.11. Continued.

Feedback

- Before starting the interview, describe your intended reader or audience to the scientist. Ask that they answer your questions in simple language that a layperson can understand.
- During the interview, if your source uses any difficult words, ask them to simplify the information. Follow-up questions such as "What does that mean?" or "How will that affect the ordinary citizen?" can help elicit a golden quote.

Learning activity 14.12a: Unsuitable quotes

Read the following hypothetical direct quote.

"Our study confirmed high levels of asymptomatic P. falciparum in the district."

Which of the following characteristics make the above quote unsuitable?

- Jargon
- Repetition
- Plain facts

Please write down your response before reading on.

Answer and feedback

- Jargon.

A good quote should be understandable and interesting to a layperson. If you have a technical term in an otherwise dramatic statement, you may insert an interpretation in parenthesis and use it as a direct quote.

Learning activity 14.12b: Unsuitable quotes

Read the following hypothetical paragraph.

"The principal investigator, Peter Balayo, said the discovery would enable households to reduce their expenditure on energy. 'People are going to save money,' he said."

Which of the following qualities make the quote unsuitable?

- Technical jargon
- Repetition
- Plain facts

Please write down your response before reading on.

Answer

- Repetition.

Quotes that merely repeat what has been stated in the preceding sentence are called "parrot quotes". They are a waste of space.

Learning activity 14.12c: Unsuitable quotes

Read the following direct quote.

"'The livestock sector in Tanzania is predominantly consisting of the traditional herd which is not very productive but very important to the livelihood of farmers and cattle keepers. Over a number of years now there has been a lot of efforts put into this sector to modernize it and improve its productivity,' the Tanzanian Minister of Agriculture, Hon. Frederick T. Sumaye, said."

(FAO, 1995)

Which of the following qualities make the quote unsuitable?

- Technical jargon
- Repetition
- Plain facts

Please write down your response before reading on.

Answer

- Plain facts.

Sentences that express insights, views, feelings and consequences tend to make better quotes than statements of plain facts. A statement of fact would have to be really dramatic to compete with opinions and feelings.

Summary

- Editors are more likely to appreciate a story on research findings if they consider it to be interesting, important and timely enough to draw the attention of their audience.
- The more relatable a story is, the higher the chances of getting published.
- You can make scientific information interesting to journalists by:
  - humanizing the story
  - relating it to a trending topic
  - relating it to people's most pressing needs.
- Human interest science stories combine human interest and science to provide interesting and socially relevant information.
- In science-related developments there are often heroes, villains, victims, survivors, unique beneficiaries and entertaining individuals whose stories can be told in the media.
- The person's story should meet one or more of the criteria for what makes news.
- Human interest stories will receive more attention if they are timed around trending discussions, events, anniversaries and policy pronouncements.
- A good direct quote is catchy and memorable.
- Strong opinions, a reference to the consequences of something, the use of colourful language, active words and dramatic expressions all make a quote memorable.
- Avoid using parrot quotes, scientific jargon and statements of plain fact as direct quotes.

References

FAO (1995) An opening speech by the Hon. Frederick T. Sumaye, (MP) Minister of Agriculture, at the FAO sponsored. In *Strategies for Market Orientation of Small Scale Milk Producers and their Organisations: Proceedings of a workshop Held at Morogoro Hotel, Morogoro, Tanzania, 20–24 March, 1995.* Available at: http://www.fao.org/3/X5661E/x5661e04.htm (accessed 1 May 2021).

Hadhazy, A. (n.d.) How has Magic Johnson survived 20 years with HIV? LiveScience.com. Available at: https://www.livescience.com/16909-magic-johnson-hiv-aids-anniversary.html (accessed 1 May 2021).

Lillian, S. (2022) Kenyan farmers suffer losses as climate change grips. SciDev.Net 31 January. Available at: https://www.scidev.net/sub-saharan-africa/feature/kenyan-farmers-suffer-losses-as-climate-change-grips/ (accessed 17 February 2022).

Ogada, J. (2019) Drug resistance could make 28 million people poor. SciDev.net 7 March. Available at: https://www.scidev.net/sub-saharan-africa/news/drug-resistance-could-make-28-million-people-poor/ (accessed 1 May 2021).

Interrogating Science

15

In this chapter
- Smelling a rat in science: how to detect suspicious research findings
- Scientific fraud
- Warning signs
- Detecting suspicious scientific claims
- Investigative tools and techniques

Smelling a rat in science: how to detect suspicious research findings

You have almost certainly heard about fake news: this thing that spreads faster than real news and that is ruffling feathers across the world. Similarly, fake science is widespread. It includes falsification of data, exaggerated interpretations of research findings or deceptive research methods. It is given different names, depending on the form it takes, such as:

- scientific hoax
- pseudo-science
- scientific misinformation
- corruption of scientific knowledge
- scientific fraud
- scientific lies
- bogus research
- sham research.

Learning activity 15.1: Should you trust it or not?
Consider whether the statements below are true or false.
1. If something is stated by a senior scientist, then it must be true. True/false?
2. If something is published in a peer-reviewed journal, then it must be true. True/false?

Please write down your response before reading on.

Answers

1. False
2. False

©2022 CAB International. Science Communication Skills for Journalists: A Resource Book for Universities in Africa (Ed. Charles Wendo)
DOI: 10.1079/9781789249675.0015

To further understand the answers to Learning activity 15.1, read the paragraph below, extracted from the article "An epidemic of false claims", which was written by John P.A. Ioannidis and published in *Scientific American* (see https://www.scientificamerican.com/article/an-epidemic-of-false-claims/).

> False positives and exaggerated results in peer-reviewed scientific studies have reached epidemic proportions in recent years. The problem is rampant in economics, the social sciences and even the natural sciences, but it is particularly egregious in biomedicine. Many studies that claim some drug or treatment is beneficial have turned out not to be true. We need only look to conflicting findings about beta-carotene, vitamin E, hormone treatments, Vioxx and Avandia. Even when effects are genuine, their true magnitude is often smaller than originally claimed.
>
> (Ioannidis, 2011)

In summary, the writer is telling us that some scientists exaggerate research results, beat the peer review system and get their dishonest or misleading papers published. Therefore, it is not true that if something is published in a journal or stated by a senior scientist it must be true. Peer review definitely helps to eliminate questionable research papers but the system is not foolproof. Some dishonest or substandard research studies slip through, especially in journals that do not follow strict editorial standards. Furthermore, most journals include sections that are not peer reviewed.

Learning activity 15.2 Motivations for false scientific claims

In your view, why would a scientist make false claims about their research?

Please write down your response before reading on.

Feedback

Scientists who publish dishonest or misleading research papers often do so to achieve a promotion or a financial reward (Ioannidis, 2011).

Scientific fraud

Whereas most scientists carry out research to generate new knowledge, a few focus on personal gain rather than scientific truth. For instance, in 2010 two Chinese scientists were sacked over alleged scientific fraud ("Chinese scientists dismissed after 70 suspect papers"), (see https://www.scidev.net/global/news/chinese-scientists-dismissed-after-70-suspect-papers/) (Ni, 2010).

Below is another excerpt from the article "An epidemic of false claims":

> Much research is conducted for reasons other than the pursuit of truth. Conflicts of interest abound, and they influence outcomes. In healthcare, research is often performed at the behest of companies that have a large financial stake in the results. Even for academics, success often hinges on publishing positive findings. The oligopoly of high-impact journals also has a distorting effect on funding, academic careers and market shares. Industry tailors research agendas to suit its needs, which also shapes academic priorities, journal revenue and even public funding.
>
> (Ioannidis, 2011)

A controversial HIV vaccine claim

In the late 1990s, a senior Nigerian doctor claimed he had invented a vaccine that could both prevent and treat HIV. Over many years, the doctor administered his "vaccine" to thousands of patients. In 2004 he published an article in the journal *Vaccine* (see https://pubmed.ncbi.nlm.

nih.gov/15364427/) (Abalaka, 2004). He cited the article as evidence of the legitimacy of his vaccine. However, the editor of the journal had published a disclaimer stating that the article was published as a report and not as a peer-reviewed research paper. Peer review gives the opportunity for a panel of scientists to vet research before it is published, and therefore reduces the risk of false claims being published. The Nigerian National Agency for Food and Drug Administration and Control (NAFDAC), the Joint United Nations AIDS Programme (UNAIDS) and the Nigerian Academy of Sciences and Network of People Living with HIV/AIDS in Nigeria (NEPWHAN) opposed the doctor's HIV vaccine claims. At one point the Nigerian government banned the treatment, saying the doctor's claims could not be scientifically verified. However, a court later nullified the ban (see https://guardian.ng/features/weekend/fresh-rumble-over-court-judgement-on-abalaka-s-hiv-vaccine-claim/) (Anuforo, 2015).

Why does it matter?

Scientific research provides the evidence that guides individuals, professionals, policymakers and other people in making decisions. But what happens if the evidence is false? Those who rely on false evidence may make a wrong, and sometimes detrimental, decision.

Learning activity 15.3: The results of false scientific claims

Can the following incidents result from dishonest or misleading research findings – yes or no?

- A doctor prescribes the wrong medication and a patient's condition gets worse or they die.
- A government approves a dangerous pesticide for use in the country.
- A journalist misinforms a million viewers, listeners or readers.
- A parent feeds a baby on an improper diet, leading to malnutrition.
- A cancer patient shifts from the prescribed hospital treatment to an unproven alternative treatment and dies prematurely.
- The public lose confidence in research.

Please write down your response before reading on.

Answers

The answer to all six is "yes".

The article "Mbeki Aids denial 'caused 300,000 deaths'" (see https://www.theguardian.com/world/2008/nov/26/aids-south-africa) (Boseley, 2008) states that, contrary to the global mainstream scientific consensus, former South Africa President Thabo Mbeki took advice from a small group of dissident scientists who wrongly claimed that HIV did not cause AIDS. As a result, the South African government was reluctant to embrace the life-saving treatment known as antiretroviral therapy. A research study carried out by a team of Harvard University scientists concluded that about 300,000 people may have died due to South Africa's delay in embracing antiretroviral therapy.

Warning signs

Now you know that false scientific claims and fake research exist, and that their impact can be tragic. Well-intentioned journalists can be duped into disseminating fake or bogus science if they do not exercise the necessary journalistic checks. So how do you avoid walking into this trap?

First, you need to know the warning signs. The first gauge is your own instinct. Is the scientist's assertion believable? Whatever feels too good to be true could well be untrue. In particular, if the assertion goes against conventional wisdom, you need to double- and triple-check it. However, many false scientific claims sound believable.

Spaghetti on trees?

In 1957, the BBC broadcast an April Fool's Day story about trees in southern Switzerland that supposedly yielded spaghetti. The story stated that an unusually light winter had given way to early flowering and bumper spaghetti harvests. It explained the process of harvesting and drying spaghetti. It concluded with a report on a traditional celebration of a spaghetti bumper harvest.

Learning activity 15.4: Spaghetti trees

Is the spaghetti tree story believable?

Please write down your response before reading on.

Feedback

"No" is the most probable answer. However, this might depend on where, and the circumstances in which, one lives. Some people believed the story and even contacted the BBC to find out where they could get spaghetti trees to grow. This is not entirely surprising, because many people eat spaghetti without knowing where it comes from and how it is made. The point here is that however implausible a scientific claim might be, there are likely to be at least some people who will believe it. A second point is that you should not believe a scientific claim simply because it sounds exciting.

Learning activity 15.5: Doubting the spaghetti story

Reflect again on the spaghetti story referred to above. What in the story would make you doubt it?

Please write down your response before reading on.

Feedback

The label on packs of spaghetti usually states that it is made from wheat. Even if you didn't know how spaghetti is made, you are unlikely to have heard of spaghetti plants before, so how could it be that farmers are getting bumper spaghetti harvests? Surely this sounds too good to be true?

The spaghetti story shows us that a good rule to follow is that if something sounds too good, or if it goes against conventional wisdom, be extra careful, otherwise you could end up believing that spaghetti grows on trees. In the case of the spaghetti story, this was just a hoax, not a misleading scientific claim, but this anecdote should make clear that what sounds too good to be true is probably not true.

Instinct aside, there are a number of things you need to check to reassure yourself of the veracity of any scientific claim.

Detecting suspicious scientific claims

Who funded the study?

Read the excerpt below from a *New York Times* article (see https://well.blogs.nytimes.com/2015/08/09/coca-cola-funds-scientists-who-shift-blame-for-obesity-away-from-bad-diets/?ref=health&_r=0):

> Coca-Cola, the world's largest producer of sugary beverages, is backing a new "science-based" solution to the obesity crisis: to maintain a healthy weight, get more exercise and worry less about cutting calories. The beverage giant has teamed up with influential scientists who are advancing this message in medical journals, at conferences and through social media. To help the scientists get the word out, Coke has provided financial and logistical support to a new nonprofit organization called the Global Energy Balance Network, which promotes the argument that weight-conscious Americans are overly fixated on how much they eat and drink while not paying enough attention to exercise.
>
> Most of the focus in the popular media and in the scientific press is, "'Oh they're eating too much, eating too much, eating too much' – blaming fast food, blaming sugary drinks and so on," the group's vice president, Steven N. Blair, an exercise scientist, says in a recent video announcing the new organization. "And there's really virtually no compelling evidence that that, in fact, is the cause."
>
> Health experts say this message is misleading and part of an effort by Coke to deflect criticism about the role sugary drinks have played in the spread of obesity and Type 2 diabetes. They contend that the company is using the new group to convince the public that physical activity can offset a bad diet despite evidence that exercise has only minimal impact on weight compared with what people consume.
>
> This clash over the science of obesity comes in a period of rising efforts to tax sugary drinks, remove them from schools and stop companies from marketing them to children. In the last two decades, consumption of full-calorie sodas by the average American has dropped by 25 percent.
>
> (O'Connor, 2015)

Learning activity 15.6: Cause for concern

Looking at the excerpt above, why are some scientists concerned about a soda company sponsoring a nutritional campaign that emphasizes exercising rather than eating fewer calories?

Please write down your response before reading on.

Feedback

Because they think the campaign might mislead the public by downplaying the role sugary drinks play in the spread of obesity and diabetes.

A soft drinks company could maintain – or even increase – its soda sales if people focused on exercising rather than reducing their calorie intake. That doesn't necessarily mean a campaign the company funds will provide false information in this regard: it only means you should be vigilant in your reporting by double-checking all of its information – for example, by getting the views of independent experts and using fact-checking websites.

Personal gain

You might have heard the humorous line that you wouldn't trust a nurse whose husband sells coffins. Of course, no nurse would kill a patient for the sake of helping their spouse to sell a coffin. Nevertheless, the joke makes a valid point about personal gain and conflicts of interest. When reporting on scientific claims, ask the question: what does this scientist or organization stand to gain from the information they are disseminating or sponsoring?

Learning activity 15.7: What would the scientist gain or lose?

Reflect again on the story of the Nigerian doctor's HIV vaccine claim. What does he stand to gain if people believe he has a safe and effective vaccine against HIV?

Please write down your response before reading on.

Feedback

He might make a lot of money!

If the information is going to help a scientist sell a product or service, it doesn't necessarily mean it's false. You simply have to be more vigilant in your reporting to establish whether the claim is true.

More warning signs

- Consider the scientist's past record: if a scientist has ever been guilty of making false claims, any information from them should be double-checked.
- Obvious falsehoods and errors: if you can identify obvious lies and errors in a report, that should cast doubt on the rest of the information.
- Where the information is published: if a scientist publishes their research findings in a magazine or newspaper and not in a peer-reviewed journal, you should smell a rat. Why would they be avoiding peer review? Similarly, if the article is published in a non-peer-reviewed section of a journal, you should ask why.

Learning activity 15.8: A hoax research paper

In 2009, a graduate student at Cornell University in New York used a computer program to generate a hoax research paper. He then submitted the paper to a peer-reviewed journal and it was accepted for publication, but he withdrew it because it was only a hoax.

What does this tell you about peer review and scientific publications?

Please write down your response before reading on.

Feedback

That the journal made such a blunder does not mean all articles published in it are doubtable. Most are truthful. However, you need to maintain a certain amount of suspicion even if an article has been published in a journal. Although peer review weeds out most of the bogus scientific claims, it is not 100 % perfect. Additionally, you need to check the credibility of the journal: for example, by checking the journal's impact factor (see https://health.library.emory.edu/writing-publishing/quality-indicators/impact-factor.html).

What to do when you smell a rat

- Ask for evidence for every claim. If the scientist gives you a journal publication as the evidence, probe further. Is the journal credible? Is the research paper peer reviewed? Do the research methods and results support the conclusion? For more guidance on interpreting research results, refer to Chapter 10.

- Every science story needs the views of credible independent scientists. This is even more crucial when dealing with a suspicious claim. National academies of sciences and universities are usually a good place to go to get credible scientists to comment on your story.
- Get the position of national and international authorities on the matter: for example, NAFDAC in Nigeria, or the WHO.
- Use fact-checking websites, such as SciCheck (https://www.factcheck.org/scicheck/).

Investigative tools and techniques

Investigative journalism defined

- Investigative journalism is the use of **innovative techniques** to unearth information of **public interest** that an individual, institution or business **wants to hide** for their own benefit.
- A typical investigative story exposes an individual, institution or business that benefits unfairly from society and that stands to lose when the facts are unearthed. Examples include a business that sells unproven technology products, a bogus journal that takes advantage of unsuspicious scientists, a government official who diverts public resources or a hospital that gives false test results in order to justify an unnecessary surgical operation.
- Credible investigative journalism plays an important role in uncovering the facts that enable individuals and policymakers to make informed decisions.

How to identify what to investigate

To identify scientific issues that are being hidden from public view and examination you need to be vigilant and watchful. The following could help you identify possible investigative story ideas.

- **Careful interrogation during interviews** – When you cover any story, pay keen attention to your interviewees. Listen carefully to their words and even the non-verbal cues they exhibit, such as facial expressions. Look at the quote below from a former correspondent for the Press Trust of India, K. Jayaraman, published by SciDev.Net (https://www.scidev.net/global/practical-guides/how-to-be-an-investigative-science-journalist-1/).

  I was once interviewing a WHO official who refused to give me further information about a mosquito research project in New Delhi, India, saying it was "sensitive" – a word that should ring alarm bells for any journalist. This led to a six-month investigation, a parliamentary inquiry and the government shutting down the project.

  (Jayaraman, 2013)

- **Tip-offs** – Sometimes you may get tipped off by an individual who has seen something that is wrong. This could be a scientist, a government official or a victim of scientific errors.
- **Controversies in science** – Something may be controversial in the eyes of the public yet accepted by science. However, if scientists – who are supposed to guide society – are the ones disagreeing among themselves then the issue may be worth investigating. If an issue is controversial among scientists, you may want to find out why. Are there hidden facts that people should know? Is it really a genuine scientific debate or do some of the scientists have hidden motives? Are there conflicts of interest?

- **Observation** – By paying attention to your surroundings, you might notice suspicious activities, posters or protest movements that could be worth investigating.
- **Social media** can help you identify complaints that can lead to a good investigative story.
- **Following news** – Keenly reading, listening to or watching the news from local and international media can help you spot facts that don't add up. This could generate potential ideas for an investigation.

How to carry out an investigation

As in any other form of journalistic newsgathering, you will be looking for information from the following categories of sources:

- human sources
- paper sources
- digital sources.

The main difference is that in investigative journalism you are trying to uncover hidden information, which requires more determination and creativity. Here is the process.

1. Preliminary research. Take time to read any reports published on the topic in other media outlets. This will show you what is already in the public domain on the topic and what isn't. Additionally, explore research studies published on your topic of investigation.

2. Discussing your idea with a colleague and editor. This will help you to decide on the scope and map out the investigation, including where to go, what questions to ask and what tools to use.

3. Data collection tools. Ensure that you have prepared and tested your reporting tools. These include audio recorders, cameras, key interview questions and writing materials, such as notebooks.

4. Formulating the right questions. Investigative journalism is rooted in finding answers to important questions. Formulate key questions on the topic that you need answers to and map out where you're going to look for the answers. For example, you might want to investigate a scientist who has put forward doubtful research but who claims that their research has been published in a scientific journal and is therefore credible. This claim should immediately raise two questions. Is it a credible journal? Was that article peer reviewed? These are questions you can answer by examining the article or journal, without interviewing people.

5. Interviews. In science investigations, you will need to interview scientists from different backgrounds on the topic you're covering. Pay close attention to your interviewees because some of their disclosures could lead to further probing. Corroboration is very important here as single-sourcing can lead to reporting inaccurate information.

6. Collecting evidence. Always ask for evidence – for example, in the form of reports, meeting minutes or travel documents – that can support you in case the person, business or institution you are exposing sues you. Remember, your exposure will cause them to lose credibility or income and they might want to punish you for it – for example, by taking you to court. This will be easier for them if you do not have evidence backing up your story.

7. Utilizing your networks. One of a science journalist's techniques – and also one of their key strengths – is networking. You should establish and cultivate relationships, especially with scientists. You will need good sources in order to obtain insider information, documents and interpretations for your investigations.

8. Data mining. This means digging through relevant datasets to establish certain patterns and to find out the reasons that lie behind these trends. Data mining requires patience: some investigations may require you to sieve through huge datasets.

9. Ethical considerations. All journalism should be carried out ethically. In conducting investigative journalism, you must bear in mind the following, in regard to ethical considerations.

- Ordinarily you should not record your interviewees without their consent. However, you may collect information undercover (for example by using hidden cameras) if it's the only possible way and the matter is of public interest.
- Always give a right of reply to those targeted by your story.
- Get permission from relevant authorities to interview children and victims.
- Report and use research evidence accurately. Avoid exaggerations.
- Only use anonymous sources when there is a compelling need to do so: for example, if their life would otherwise be at risk. Even then, get the source to provide you with some form of evidence for their claims: for example, documents or photos.
- Share ethical dilemmas with your editor. Editors and colleagues can help you make the right decision.

What next?

- Seek legal advice on issues in your story that could raise legal concerns. This includes issues such as defamation, the right to privacy, libel and slander. This can vary from country to country. You could get a lawyer to review your investigative story before publication to avoid legal problems.
- Keep good records of your interviews and all forms of evidence you gather.
- Do follow-up stories.
- Prepare for conflicts. If you reveal what people want to remain hidden, there are likely to be conflicts of some kind. Journalists have faced court battles, death threats and even death as a result of the stories they have produced.

Summary

- Not every claim made by a scientist is truthful.
- Not every paper published in a journal is truthful.
- If it sounds too good to be true, it's probably false.
- By effectively using journalistic skills one can detect and reject fake science.
- Investigative journalism involves uncovering the issues in science that others want to hide, usually those in power or business.
- Investigative journalism involves critical analysis of facts and should be of public interest.
- Investigative journalism goes beyond an announcement of a new scientific initiative, and includes an analysis of the underlying facts. You have to ask yourself: "What do people not know that they should?"

References

Abalaka, J.O.A. (2004) Attempts to cure and prevent HIV/AIDS in central Nigeria between 1997 and 2002: opening a way to a vaccine-based solution to the problem? *Vaccine* 22, 3819–28.

Anuforo, E. (2015) Fresh rumble over court judgement on Abalaka's HIV vaccine claim. Guardian.ng 12 February. Available at: https://guardian.ng/features/weekend/fresh-rumble-over-court-judgement-on-abalaka-s-hiv-vaccine-claim/ (accessed 3 May 2021).

Boseley, S. (2008) Mbeki Aids denial "caused 300,000 deaths". TheGuardian.com, 26 November. Available at: https://www.theguardian.com/world/2008/nov/26/aids-south-africa (accessed 3 May 2021).

Ioannidis, J. (2011) An epidemic of false claims: Competition and conflicts of interest distort too many medical findings. ScientificAmerican.com 1 June. Available at: https://www.scientificamerican.com/article/an-epidemic-of-false-claims/ (accessed 3 May 2021).

Jayaraman, K.S. (2013) How to be an investigative science journalist. SciDev.net 25 March. Available at: https://www.scidev.net/global/practical-guides/how-to-be-an-investigative-science-journalist-1/ (accessed 3 May 2021).

Ni, W. (2010) Chinese scientists dismissed after 70 suspect papers. SciDev.net 13 January. Available at: https://www.scidev.net/global/news/chinese-scientists-dismissed-after-70-suspect-papers/ (accessed 3 May 2021).

O'Connor (2015) Coca-Cola funds scientists who shift blame for obesity away from bad diets. Well.blogs.nytimes.com 9 August. Available at: https://archive.nytimes.com/well.blogs.nytimes.com/2015/08/09/coca-cola-funds-scientists-who-shift-blame-for-obesity-away-from-bad-diets/?ref=health&_r=0 (accessed 22 July 2022).

Earning a Living from Science Journalism

How to make a compelling pitch

Pitching a story idea is all about convincing an editor that your story idea is interesting, important and timely enough for them to take it on. It requires an understanding of what makes news in science, but that is not all. A journalist needs to know how to make their story idea so compelling that the editor will prioritize it over others. And, finally, a journalist needs to have the right techniques for presenting their story idea to an editor.

In Chapter 13 we discussed what makes news in science, and additional techniques for making scientific information interesting to non-specialists. In this chapter we shall look at how to use those principles to convince editors and givers of reporting grants that your story idea is worth selecting.

Techniques for pitching to an editor

In pitching a story idea you are attempting to quickly catch the editor's attention and convince them that you have a newsworthy story. But first it's important to know the circumstances in which editors operate.

©2022 CAB International. Science Communication Skills for Journalists: A Resource Book
for Universities in Africa (Ed. Charles Wendo)
DOI: 10.1079/9781789249675.0016

Learning activity 16.1: Editors

Write down the following points about the circumstances in which editors work. For each of them, consider what this means for the way you should present your pitch.

- Editors are busy people. They may not have a lot of time to listen to you or to read your pitch.
- Editors are constantly under pressure to grow their audiences.
- At any one time, various people and organizations are competing for the editor's attention.

Please write down your response before reading on.

Feedback

Editor's circumstances	What it means for you
Editors are busy people. They may not have a lot of time to listen to you or to read your pitch	Aim to impress them within the first one to three sentences. An email pitch should have a subject line and first sentence that are interesting enough to draw a busy editor's attention
Editors are constantly under pressure to grow their audiences	You need to convince them that the information you are providing will be attractive to their audiences
At any one time, various people and organizations are competing for the editor's attention	To be successful in pitching you need to stand out from the crowd by providing compelling information.

Pitching guidelines and templates vary from one media organization to another but they have the following elements in common.

- What is the story?
- Why does it matter?
- Why now?
- How do you plan to approach the story?

What is the story?

In one clear sentence, you should be able to tell an editor what the story is about. It should be clear that your story is going to reveal something new. It could be a new research finding, an event, a trend, an important pronouncement, a controversial viewpoint, etc. Whatever it is, you should be able to state it in one sentence.

Why does it matter?

In other words, what is at stake? Why should anybody care? How does the new development affect citizens, and who is affected? What are the implications for society?

Why now?

Is there any urgency? Why must the story be told now and not later? It could be that this issue is on people's minds right now, and therefore the story is timely. Or there could be a lot to lose if the public doesn't get to know about it immediately. It may also be something that society has been eagerly waiting for. Or it might be a new research study that has just been published in a journal.

How do you plan to approach the story?

If your first line has impressed the editor, they will read on or listen further to find out how you will approach the story. What kind of information will you collect? Where will you get this information? What aspects of this issue would people be most interested in hearing about, and why? Who are the people you plan to interview? What questions will you ask them? What kind of documents or online resources will you need to look at?

Example

What is the story? A paralysed man has been able to walk again after a ground-breaking surgical operation.

Why should anyone care? There are millions of paralysed people out there who could benefit from this kind of operation. This story will provide hope to them and their families.

Why now? The information has just been released.

How to win and use a story grant

Journalists can often find free money to help them to report science stories that will benefit society. You need to know where to find that money and how to apply for it.

Identifying the opportunity

Givers of story grants usually make announcements according to their calendars. The trick is to make sure you don't miss these announcements. Below are a few ways to position yourself to ensure you don't miss grant announcements.

1. Join professional associations, such as the World Federation of Science Journalists (WFSJ). As at April 2021, WFSJ had over 67 member associations across the globe, representing over 10,000 journalists. These members get regular updates from WFSJ, including tips, professional development opportunities and story grants. Find and join a science-related journalism association within your country, especially one that is a member of WFSJ.

As at July 2022, the following Africa-based associations were members of WFSJ:
- African Federation of Science Journalists (AFSJ)
- South African Science Journalists' Association (SASJA)
- Arab Science Journalists' Association (ASJA) - based in Egypt
- Association des Journalistes et Communicateurs Scientifiques du Benin (AJCSB)
- Association des Journalistes et Communicateurs Scientifiques du Burkina Faso (AJSC/BF)
- Association des journalistes scientifiques et communicateurs pour la promotion de la santé (AJC-PROSANTE) – Cameroon SciLife (Cameroon's Association of Science Journalists and Communicators)
- Association Congolaise des Journalistes Communicateurs Scientifiques (ACJCS) (Democratic Republic of Congo)
- Medical Journalists' Association of Ghana (MJAG)
- Association des Journalistes Scientifiques de Guinée (AJSG)
- The Kenya Environment and Science Journalists' Association (KENSJA)
- Médias Pour la Science et le Développement (MSD) – Ivory Coast Media for Environment, Science, Health and Agriculture (MESHA, Kenya)
- Association des Journalistes Scientifiques du Niger (AJSN)
- Nigeria Association of Science Journalists (NASJ)
- Rwanda Association of Science Journalists (RASJ)
- Somali Media for Environment, Science, Health and Agriculture (SOMESHA)
- Science Journalists and Communicators of Togo (JCS-Togo)
- Uganda Science Journalists' Association (USJA)
- Zimbabwe Environmental Journalists' Association (ZEJA) Zimbabwe National Association of Science Journalists (ZNASiJ)

2. **Subscribe to the mailing lists and social media accounts of grant-giving organizations**, such as the African Academy of Sciences (https://www.aasciences.africa/aesa/programmes/science-communicationafrica-science-desk-asd) and the Bill and Melinda Gates Foundation (https://www.gatesfoundation.org/).
3. **Use online resources, such as the following:**
   - the International Journalists' Network opportunities page (https://ijnet.org/en/opportunities)
   - the Global Investigative Journalism Network grants and fellowships page (https://gijn.org/grants-and-fellowships-2/), which has a list of grant and fellowship opportunities targeted to investigative journalists
   - the Fund for Investigative Journalism (http://fij.org), which has story grants for various topics, including climate change, that can be accessed by journalists globally.

Please note that the list of organizations that give grants changes from time to time. Similarly the sources of information on these grants keep evolving. Therefore, you need to be on the lookout for the latest information on reporting grants, for example by paying attention to the relevant newsletters.

What makes a good grant proposal?

Grant applications are highly competitive. To win, you need to write a compelling application. Below are some of the characteristics of a good application for a story grant.

Learning activity 16.2: Stating your story idea in a grant application

Which of the following statements will be more appealing to the jury who decides on the approval of a grant?

Statement 1: "My story is about Fall Armyworm in Ethiopia"

Statement 2: "My story is about how biopesticides are helping farmers to control Fall Armyworm in Ethiopia"

Please write down your response before reading on.

Feedback

Statement 2 would be more appealing to the jury because it is more specific and is clearer.

- **Clarity** – Your proposal should be clear, specific and easy to understand. The jury should not struggle to understand your proposal. Pitch a specific story idea and not a broad theme.
- **Clearly state how your story is likely to contribute to efforts to solve a specific problem** – Might your story lead to a much-needed policy change? Can it inspire farmers to adopt a better farming method? Is it going to draw attention to a problem that needs to be resolved?
- **Read the terms and conditions of the grant well** – Take note of who qualifies for the grant, countries where the grant can be used, the kind of stories that the grant can be used for, and the application procedure.
- **Beat the deadline** – All grant applications have timelines. Ensure that your application is submitted by the set deadline. Some grant administrators have a policy of rewarding early applicants by selecting them over late applicants if they score the same points on other issues.
- **Demonstrate potential for success** – Grant administrators want to be persuaded that the story is doable and that you are the right person to do it. It is important to cite other stories you have published.

- **Have a well-organized activities plan in place** – A good activity plan will help the grant administrators understand the details of what you intend to do, at what time and in what location, and that this meets their requirements.
- **Seek guidance** – If you need any clarification, contact the grant administrators directly for guidance. Additionally, some organizations have a frequently asked questions (FAQs) section that might have answers to some of your questions.
- **Have a realistic budget** – Ensure that your budget estimates are within the standard pricing limits. Juries need to have a snapshot of the costs you are likely to incur. It can be difficult to list all the unforeseen costs but do not make budget proposals that are excessive.
- **Always proofread** – Before submitting an application you should have it proofread: grammatically incorrect proposals can put off the jury as they may struggle to understand your pitch.

How to use a story grant

One of the challenges that reporting grants can pose to journalists is remaining objective when using the grant. It is important to remain objective even when you are benefiting from a story grant. Acting independently is one of the key principles of journalism ethics. If the grant requires you to promote the specific interests of an organization against the interests of society, that will be in conflict with the normal ethics of journalism.

It is advisable to disclose that your story was funded through a grant, and to name the awarding organization. Disclosure is necessary, especially when there are potential conflicts of interest, or when the circumstances could influence your coverage.

Summary

- To make a compelling pitch it is important for a journalist to use the techniques for making a science story interesting to non-specialists.
- When pitching to an editor, it is important to begin with one clear sentence that catches their attention.
- A pitch should clearly state what the story is, the implications, the urgency and how you intend to approach it, including time frame and budget.
- To increase their chances of finding story grant opportunities it is important for a journalist to join science-related media associations, subscribe to mailing lists and social media accounts of potential grant-giving organizations and use online resources.
- To increase their chances of winning a story grant, journalists need to:
  - write a clear and specific proposal
  - state how their story is likely to solve a specific societal problem
  - read the terms and conditions of the grant
  - meet the deadline
  - show that the story is doable and that they are the right person to do it
  - have a good activity plan in place
  - consult the grant administrators for clarification if necessary
  - have a realistic budget
  - proofread the application before submitting it.
- When using a story grant, it is important for journalists to remain independent and objective.

Index

Note: Page numbers in **bold** type refer to **figures** and page numbers in *italic* type refer to *tables*